大廚小菜

三姐的50道創意家常菜

（增訂版）

Master chef's home-style favourites (revised edition)

50 creative recipes by Chef Kitty Siu

蕭秀香（三姐）著

Preface

推薦序

小菜不簡單

許多人或許只知三姐蕭秀香是電視烹飪節目主持人，但在我眼中，她更加是女中豪傑，在飲食界早已打出名堂，曾在新城電台主辦的「食神爭霸戰」中贏過雙冠軍和最具創意獎。

不要以為這是簡單的事，在傳統華人飲食界中，女廚師的地位一向不高，以為她們沒有能力擔大旗；但三姐卻一枝獨秀，突破了這個舊觀念，打理兩間食肆時，充分表現其膽大心細的風格，令一眾男廚師都甘拜下風。

三姐在我一個電台節目上曾分享，原來她從小師承父親。家中排行中間，只有她一個願意在熱廚房幫手，就是這樣練成一身功夫，擅長就地取材，把不同而合時節的食材，靈活配搭，煮成多道好味而新穎的菜式。

難得她今次慷慨分享多年來的心得，把精心製作具有創意的家常小菜公諸同好。不要輕看這四十多道的小菜，如非有大廚的歷練，也煮不出來。

張宇人

立法會議員（飲食界）

急智廚王三姐

香港粵菜廚師人才濟濟，但要數
粵菜女廚師，第一個想起的，非三姐莫屬。

三姐廚藝高、人疏爽，做節目每每做一餐餸，可能「筆直」是一百蚊一餐，三姐總會自掏腰包買夠二百；過時過節，買夠三百。為的就是想大家開心，讓眾人能享受到美食，真的，大家都因享受到三姐的廚藝而喜孜孜。三姐煮餸，功架十足，氣勢非凡，眼神凌厲；令所有人都完全感受到她烹調的樂趣。家常便飯，三姐都可變成豪門夜宴。我們五花八門的問題，三姐隨聲應對，三言兩語，回應得綽綽有餘。陪三姐煮餸，腦筋轉慢點都「死火」。

三姐——人未見，聲先到；人開心，煮的菜也令人吃得快樂，吃得笑逐顏開。這一位出色的女廚師、廚藝界殿堂級大師，將愛心、開心融會在刀下鑊裏，寫成這本家常小菜送給大家。一字記之曰：「正！」

安德尊

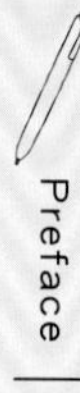

日日創新　永不停步

數一數手指，原來認識「三姐」已經二十多年了。她的精湛廚藝早已降服了一眾老饕，就算我再有精彩的描述，也只不過是錦上添花罷了！所以，我打算從其他角度去介紹「三姐」，讓讀者們可以更全面了解這位隱世女廚神。

所謂大隱隱於市，「三姐」從來不慕名利，一心只想將承襲自父親的家業做好，在傳統浩瀚的烹飪功夫上打下扎實的基礎。但她深明食家的要求是永無止境；所以，無論在食材方面與及烹調手法均能推陳出新，再配以各地飲食文化的精髓。如今的「三姐」已經脱胎換骨，卓然成家。

其實「三姐」除了廚藝了得，她的口才也不弱，從電視上見她與主持人的對答，字字珠璣，加上笑容可掬，平易近人。殿堂級廚神而無架子，實屬難得！

自古及今，廚神多矣！商湯時有伊尹，齊桓公有易牙，其後有庖丁諸人，似乎鼎鑊之事均操於男子之手，鮮有女子出頭；而今「三姐」能夠並肩其間，亦為飲食界增添色彩。

今聞「三姐」將平生心得結集成書，我等企望久矣！有幸先睹為快，讀之已覺香味飄然而至，食指大動。

林綺嫻

工商管理學博士
香港餐務管理協會顧問

三姐，既是名廚，也是著名廚藝專家、電視節目主持人，

還是一位善心人，經常參與各類慈善活動。

Foreword

自序

烹飪，一直是我最喜歡的興趣。

幾十年來，經過我處理的食材及調味品數之不盡，也烹調過數不清的菜餚，自問對每樣食材的特性及味道也「知根知柢」。材料總離不開豬、牛、雞、魚、菇、豆類等等，烹調是讓你在平凡的食材之中帶出不平凡的菜式來，烹調的人要付出百份百的心思，調節出新口味、口感獨特的菜餚。說真的，我就是喜歡千變萬化的烹調。

2019 年出版的《大廚小菜》有幸獲得讀者支持，今次推出增訂版本，我加添了五個全新食譜，以及乾貨處理技巧、急凍肉處理等，希望帶給讀者全新的烹飪靈感，加上自己的無窮創意，將書本內的靈感加以發揮，創作別樹一幟的自家作品，家人滿意之餘，自己也開心，自有一份滿足感！

食譜書是啟發烹調靈感之源，期望大家從書本找到入廚的點子發揮創意，不斷提升煮餸興趣，將食材加以變化，創作更多令自己及家人滿意的新菜式。

三姐

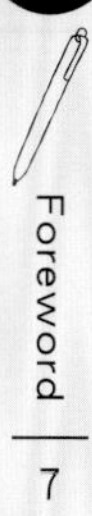

Contents

目錄

火鴨絲梅子蝦 81
Fried prawns and shredded roast duck in plum sauce

特色小炒王 84
Assorted stir-fry with dried and fresh seafood

魚類 Fish

鴛鴦蘿蔔燜魚天翅 88
Braised fish fin with fresh and salted radish

梅子半煎煮黃花 92
Braised yellow croaker in lemon plum sauce

薑葱燜魚卜魚雲 96
Braised fish head and swim bladders with ginger and spring onion

脆燜斑頭腩 100
Braised grouper head and belly

吊燒脆鱔（三姐示範） 104
Fried crispy eel

橙汁香芒炒鱔球 107
Stir-fried eel in orange mango sauce

海參魚蓉粥 110
Sea cucumber and fish fillet congee

粟米金銀魚燴飯 113
Grouper fillet and salted fish fried rice in sweet corn sauce

肉類 Meat

脆骨蒸肉餅 116
Steamed pork patty with diced pig ear

蜜桃咖啡肉排 119
Coffee-scented pork ribs with peaches

欖菜豆豉骨 122
Pork ribs with pickled olive mustard greens and black bean sauce

酸汁凍牛腒伴青瓜 126
Beef shin cold appetizer with cucumber in Sichuan vinaigrette

香葱肉丸蛋中蛋 130
Scrambled egg with pork balls, egg white omelette and spring onion

番薯咖喱牛腩 133
Beef brisket curry with sweet potato

芥蘭頭年糕牛柳粒 138
Seared beef tenderloin cubes with kohlrabi and glutinous rice cake

蜜汁欖角骨 141
Pork spareribs in honey olive sauce

涼瓜酸薑炒牛柳 144
Stir-fried beef tenderloin with bitter melon and pickled ginger

家禽 Poultry

薑葱蜆蚧雞煲 148
Chicken casserole with ginger, spring onion and fermented clams

檸檬冬蔭雞翼 151
Tom Yum lemon chicken wings

芝麻乳鴿 154
Roast sesame squabs

菜蔬 Vegetable

鯪魚肉韭菜煎藕餅 158
Minced dace patty with chives and lotus root

煎釀秋葵 162
Pan-fried stuffed okras

XO 醬肉碎炒花生芽 165
Stir-fried peanut sprouts with ground pork in XO sauce

薑蓉蒸勝瓜腐皮卷 170
Steamed beancurd skin roll and angled luffa with ginger and garlic

尊貴素薑醋 173
Assorted vegetables with ginger in sweet vinegar

鍋貼小棠菜 178
Deep-fried minced cuttlefish patty with Shanghainese Bok Choy

銀魚乾魚香蒸茄子 182
Steamed eggplant with dried anchovies and chilli black bean sauce

雜菇蛋包年糕 185
Assorted mushroom and rice cake omelette

蜆肉薯蓉餅 188
Potato clam croquettes

雪菜金菇煎豆腐 192
Pan-fried tofu with Xue Cai and enokitake mushrooms

袋袋平安 196
Beancurd skin beggar's purse

Kitty's secret tips

三姐
入廚小秘技

輕鬆炮製美味佳餚

烹調任何菜式前，妥善處理食材是關鍵的第一步。對於新手入廚人士來説，往往摸不着頭腦，究竟該如何入手準備食材？

今次，我從乾貨海味、牛肉調味等説起，掌握食材的特點從而炮製出色的菜餚，與家人共享美食。

除此之外，如何控制火候？ 怎樣炸出外脆內嫩的炸物？ 在家煎炒炸時，如何節省用油量？

此書介紹的全是在家容易炮製又有創意的菜式，可用家庭式廚具煮出來，大家一起跟着做，為家人炮製溫暖的美食吧！

乾貨海味的處理方法

不少菜式會配搭海味乾貨烹調，如香氣濃郁的冬菇、鮮甜的乾瑤柱、滋補的海參等都是家常菜不可或缺的食材，我自己也經常在菜譜使用，事前有哪些步驟需要注意呢？

乾冬菇

- 我喜歡選用花吋菇，吃入口時一啖一隻，口感豐富、軟滑。
- 使用前必須用水浸軟，肉質軟腍才能更易入味，所以必須預留足夠浸泡時間。
- 冬菇泡軟後，與糖、生粉及油 1 湯匙拌勻，再放入薑及紹酒蒸 2 小時，待涼後可分成小包儲存於冰箱，使用時取出解凍即可，可存放 4 個月。
- 開啟包裝的冬菇宜放於雪櫃冷藏，免受害蟲蛀食。

乾瑤柱

- 乾瑤柱洗淨後，先用水浸 5 分鐘，水倒掉，再用水浸泡至軟身，泡浸水不要棄掉，可留作烹煮之用，令菜式更鮮甜好味。
- 乾瑤柱除了泡至軟身之外，可加少許水蒸腍，撕成幼絲備用。
- 蒸至軟腍或炸透的乾瑤柱可存放密實袋，冷藏冰箱備用。
- 大粒的乾瑤柱適合用於原粒瑤柱甫，宜選購赤色的，味道較甜；如大粒碎狀的乾瑤柱，也可用於炒飯、炒粉絲或煲湯等。
- 乾瑤柱旁邊有一個枕，一般家庭烹調可不用去掉，或切出用於熬湯，鮮味濃郁，以免浪費食材。
- 我習慣將胡椒粒用茶袋包好，與乾瑤柱一起擺放，以防蟲蛀。

海參

- 海參要經過反覆換水及煲煮焗透等程序，可將泡浸的海參置於雪櫃待軟，可減低壞掉的機會。
- 海參泡至軟身後，去掉內臟，用手指捽洗乾淨，再用薑葱煨煮，可隨時取用。
- 浸泡海參的器皿切勿沾有油分，否則海參溶掉不成型。

陳皮

- 柑皮儲藏三年以上稱為陳皮，我通常選用五、六年以上的陳皮來蒸肉餅、蒸魚或煲湯，氣味淳香，令菜式回味無窮。
- 陳皮泡軟後，如用於蒸魚，我會刮淨內瓤再切成幼條，蒸出來的魚更晶瑩及美觀，口感也較佳。

牛肉調味有先後

烹調牛肉前，我習慣先放入糖、油及生粉醃勻牛肉，先封住牛肉表面肉質，令牛肉的肉質軟滑，隨後才下生抽加味，令鹹味不會太濃。

煮出好味急凍肉

這陣子很多人習慣購買急凍肉儲糧，其實烹調急凍肉也可以煮出好味餸菜，秘訣在於解凍得宜。急凍肉解凍時要徹底啤水，將肉質回軟過來，待完全沒有血水滲出後，用廚房紙吸乾水分，再加調味料入味即可。不過，牛扒則不適用於這個解凍方法。

我家的常備料

我曾經是 OL，知道下班後為家人快手煮一餐不是易事。在這裏與大家分享一下我家廚房的常備料，希望能為你帶來煮食靈感。

- 牛肉、豬扒
- 瓜、番茄
- 雞蛋、鹹蛋
- 薑*、蒜頭、紅葱頭
- 蝦米、蝦乾、瑤柱、八爪魚乾、小銀魚乾、乾魷筒
- 冬菇、雲耳、粉絲
- 梅子、梅菜
- 蜆蚧醬、麵豉醬、蝦醬、蠔油、XO 醬

* 我喜歡用小肉薑，薑味香濃，而且帶皮使用。薑連皮食用很有好處，薑皮既有營養，而且性涼，與熱性的薑肉一起剛好保持生薑藥性的平衡。街市常見的大肉薑很多已浸水，味道不好且易爛，選購時一定要揀乾身的。

30 分鐘三餸一湯

煮一餐也如是。有些東西必須即煮即食，有些煮好放一會也沒關係，不會影響食味，有些甚至要放涼才好吃；所以煮時要分先後。就以三餸一湯的家常菜來說，既要快手也要樣樣好吃，烹製方法就是關鍵。三個餸最好蒸、煎或焖、及小炒，次序是先煮飯和煲湯，同時煲滾水準備蒸食材，然後起鑊煎煮或爆香材料焖煮，將近開飯才炒菜。拿捏好這個原則，30 分鐘開飯絕對不是難事。

好好掌握火候

不管做甚麼菜，「火路」是最重要的。必須按照食材的質地，把握好火力的大小和加熱的時間，才能煮出合意的口感。因為每種食材耐熱程度不同，要炒好一碟餸菜，必須注意下材料的次序，先硬後軟，有些還要逐樣分開處理，千萬不要全部一次下鑊。我的食譜裏，都已經把下料的次序寫得清清楚楚。

- 薑比蒜頭耐火，起鑊時先把薑爆香之後，才加蒜頭和紅葱頭。
- 硬蔬菜如四季豆，炒之前必須走油，炸至半熟；燈籠椒可走油或用熱水浸泡；才與其他易熟材料如海鮮或肉類一齊炒熟，又能保持爽脆的口感。
- 軟的水果，如蜜桃、芒果、士多啤梨等，在炒其他材料前先用熱油或熱水浸泡，待其他材料炒熟，再放入鑊中，輕輕炒幾下即可上碟。這樣的水果裏外俱熱，不會因為煮的時間短而「面熱心冷」。

不同的食材對火候有不同的要求：

- 海鮮肉質細嫩，適宜大火速蒸或快炒。
- 煎魚時，要先用大火燒熱鑊和油才將魚放下去，煎一會調至細火慢煎，魚就會皮脆肉嫩。
- 雞鴨、豬肉等肉質較結實，需要猛火長蒸，或切片或絲快炒。
- 豬扒、牛扒要先用大火煎，鎖緊表面，然後轉中火慢煎至中間漸熟；豬扒必須熟透，牛扒則按各人口味煎至不同熟度。
- 對易熟細嫩食材，如蒸雞蛋、豆腐類要用中小火，蒸的時間也不可太長。

外脆內嫩的香炸竅門

炸物酥脆香口，製作時間又短，是受歡迎的烹飪法。

炸，通常用中大火將油燒至熱度足夠才放入食材，如油不夠熱，炸的東西表面不能迅速脫水，油會不停滲入食材，表皮軟綿綿，而且很膩，這就很難補救了。

通常直接炸熟的食材，在油剛冒煙時放入，這時的油大約五成熱（約150℃）。炸至微微金黃色並浮起，表示差不多熟了，可盛起；開大火把油溫加熱至七成（約200℃），放入食材翻炸幾秒即撈出，可達到皮脆肉嫩的口感。炸食物千萬別過火，因等候翻炸時及翻炸後，食物吸收了的熱力會漸漸逼入食物中心，內層變熟，表面也會變得更焦香。

*小秘技：

如用了較多油炸魚或海鮮，炸完後放一塊薑，可去除腥氣，並可把油留作熟油之用。

省油兩招：半煎炸及側鑊法

家庭烹飪不會用很多油，爐火也不會很猛，所以很多朋友老是說家中做的菜不夠香。只要方法用得對，在家一樣做出香氣四溢的惹味菜式。

煎炒炸一定要用油；為了不浪費，我的食譜很少建議用大量油炸食物，其實很多炸的步驟，可以用半煎炸的方式代替，只需用約 3 湯匙油就可了。

- 半煎炸排骨、魚、墨魚餅、蝦餅等，建議沾粉（炸粉或生粉）和蛋，至於是乾粉還是粉漿要視乎要求的效果而定。粉漿有兩項功能：第一，有助保持肉汁，令成品外脆內軟；其次，增加金黃香脆的味道和口感。

- 油的功能是傳熱，用平底鑊給食材走油，只要將鑊輕輕傾側，油自然集中在鑊邊，火力也較集中，炸東西無難度，也不必為少許食材而用上半斤油。

平底鑊及筷子的妙用

現在很多家庭沒有中式圓鑊，其實平底鑊也很好用，我自己也經常用來煮食。

- 上面提及的側鑊法，除了可慳油外，還可以同時在一隻鑊用大火和慢火煮東西，例如炒豬肉和爆蒜蓉，先下油把豬肉炒幾下，將平底鑊打側，豬肉在近火那邊繼續炒，蒜蓉則在離火較遠的一邊，就如用慢火爆香，不怕變焦。

- 炒餸除了鑊鏟外，一雙長筷子也是必要的工具，尤其要把食材炒散，筷子比鑊鏟更方便、更得心應手。煎魚或其他較大的東西，反轉時也必須用筷子幫忙。

經常有人問：「我明明照住食譜做，點解總是煮得不好食？」食譜固然可以幫你了解怎樣做好菜，是入廚好幫手，但更重要的是不斷實踐，在失敗中累積經驗，多花點心思解決和改進，一定會愈煮愈好；我的入廚心得也是這樣一點一滴累積起來。至於口味，人人不同，鹹淡甜辣要自己把握調整了。

準備時間 10 分鐘

烹製時間 30 分鐘

罐頭鮑魚燜釀豆腐

Stuffed tofu with canned abalone

材料

罐頭鮑魚 1 罐
節瓜 1 個
免治豬肉 1/2 斤
墨魚膠 4 兩
硬豆腐 2 件
葱 2 條，切粒
水適量

調味料

鹽 1 茶匙
生粉 1 茶匙
糖 1/2 茶匙

準備

1. 節瓜刮去皮，切件。
2. 每件豆腐切開 4 件，備用。

做法

1. 將肉碎、墨魚膠放入大碗，加入調味料及水調勻，用手撻至起膠。
2. 每件豆腐釀入肉碎墨魚膠，備用。
3. 燒熱油鍋，放入釀豆腐煎香取出。
4. 瓦煲內加入節瓜件，倒入鮑魚汁焖 10 分鐘，再加入釀豆腐及鮑魚焖 5 分鐘，最後以少許生粉水埋芡即可。

三姐心得

- 罐裝鮑魚汁鮮香惹味，丟掉的話很浪費，不妨物盡其用，配搭其他食材焖煮，善用每樣材料之餘，也能創作另一道新口味佳餚供家人品嘗。
- 節瓜以生抽調味會令味道變酸，建議用鮑魚汁或蠔油調味。

新菜式

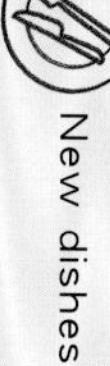

Stuffed tofu with canned abalone

- Preparation time: 10 minutes
- Cooking time: 30 minutes

Ingredients

1 can abalone
1 Chinese marrow
300 g ground pork
150 g minced cuttlefish
2 pieces firm tofu
2 sprigs spring onion (diced)
water

Seasoning

1 tsp salt
1 tsp caltrop starch
1/2 tsp sugar

Preparations

1. Scrape off the skin of the Chinese marrow with a metal spoon. Cut into chunks.
2. Cut each piece of tofu into quarters. Set aside.

Method

1. Put ground pork and minced cuttlefish into a mixing bowl. Add seasoning and water. Mix well. Lift the mixture off the bowl and slap it back hard into the bowl repeatedly under the mixture turns sticky.
2. Stuff each piece of tofu with the pork-cuttlefish mixture. Set aside.
3. Heat oil in a wok. Put the stuffed tofu in with the pork-cuttlefish filling side down first. Fry until all sides golden. Set aside.
4. Put in the Chinese marrow and pour in the brine from the canned abalone. Turn the heat down and simmer for 10 minutes. Put in the stuffed tofu and the abalone. Cook for 5 minutes. Stir in some caltrop starch slurry to thicken the sauce. Serve.

Kitty's cooking tips

- The brine that comes in the can with the abalone is full of umami and flavours. It would be too wasteful to just pour it down the drain. Feel free to use it like a stock to cook other ingredients in stews and simmered dishes. Not only can you make good use of the brine, but also create new exciting flavours for your family.
- Chinese marrow would turn sour if seasoned with soy sauce. I prefer cooking Chinese marrow with oyster sauce or abalone sauce instead.

Shrimps in zesty young ginger sauce

子薑蝦

準備時間 10分鐘

烹製時間 15分鐘

材料

蝦肉 1/2 斤
酸子薑 2 兩
三色椒適量
杏仁片適量

醃料

雞蛋白 1 個
生粉 2 湯匙

汁料

日本酸醋 2 湯匙
味醂 1 湯匙
鎮江醋 2 茶匙
酸梅 1 粒，剁碎
鹽 1/2 茶匙
糖 1 茶匙

準備

1 蝦肉洗淨，切雙飛狀，用蛋白及生粉拌勻。
2 酸子薑及三色椒切絲。
3 汁料拌勻備用。
4 杏仁片略炸至微黃色，盛起，瀝乾油分。

做法

1 燒滾油，放入蝦肉炸至微黃色，盛起。
2 另起油鑊，下三色椒走油，盛起。倒去多餘油分，下汁料用小火煮熱，以生粉水埋芡。
3 最後拌入蝦球、酸子薑及三色椒，灑上杏仁片即可。

■ 蝦肉走油時要快速，至七成熟（即轉至微黃色及粉漿凝固），待稍後蝦球回鑊時口感剛好，否則肉質變韌。

Shrimps in zesty young ginger sauce

- Preparation time: 10 minutes
- Cooking time: 15 minutes

Ingredients

300 g shelled shrimps
75 g pickled young ginger
red, yellow and green bell peppers
almond flakes

Marinade

1 egg white
2 tbsp caltrop starch

Sauce

2 tbsp Japanese ponzu sauce
1 tbsp mirin (Japanese sweet cooking wine)
2 tsp Zhenjiang black vinegar
1 pickled sour plum (de-seeded, finely chopped)
1/2 tsp salt
1 tsp sugar

Preparations

1. Rinse the shrimps. Butterfly along the back. Add egg white and caltrop starch. Mix well.
2. Finely shred the pickled young ginger and bell peppers.
3. Mix the sauce ingredients well. Set aside.
4. Deep-fry the almond flakes until lightly browned. Drain well.

Method

1. Heat enough oil in a wok for deep-frying. Deep-fry the shrimps until lightly browned. Drain and set aside.
2. In another wok, stir-fry the bell peppers in oil briefly. Set aside. Drain most of the oil in the wok. Pour in the sauce ingredients. Cook over low heat until it boils. Stir in caltrop starch slurry. Cook until it thickens.
3. Put the fried shrimps back in the sauce. Add young pickled ginger and bell peppers. Toss well to coat evenly in the sauce. Sprinkle with almond flakes on top. Serve.

Kitty's cooking tips

- When you deep-fry the shrimps, make sure you do it quickly. Fry them till the crust firms up and turns light golden (about medium-well done). Do not cook them through the first time around. Otherwise, they will overcook and turn rubbery when you toss them in the sauce at last.

Sesame-crusted crispy chicken wings

芝麻脆皮雞翼

材料 

雞翼 10 隻
蛋白 2 個
白芝麻適量

醃料 

鹽 1 茶匙
薑汁 1 茶匙

雞皮水 

麥芽糖 3 湯匙
浙醋 3 湯匙
紹酒 2 湯匙
水 1 杯

準備 

1. 雞翼浸泡水至解凍，抹乾水分，用醃料醃約 2 小時，汆水備用。
2. 雞皮水拌勻，座於熱水內至糖溶。
3. 雞翼放入雞皮水內拌勻，放在竹網上吹乾約 4 小時，塗上蛋白，灑上白芝麻，再吹乾 1 小時。

做法

1. 燒熱油，待油溫剛熱時，放入筷子見出現少許泡沫，放入雞翼以慢火炸至金黃色。
2. 上碟前，調至大火再炸一會即可。

三姐心得

- 雞翼第一次下油鍋炸時，注意油溫勿過高，否則雞翼外皮會短時間焦黑，影響賣相。
- 雞翼上碟前，調至大火再炸一會，目的是讓雞翼的油分逼出，吃時不會滿口油膩。

Sesame-crusted crispy chicken wings

- Preparation time: 7 hours
- Cooking time: 10 minutes

Ingredients

10 chicken wings
2 egg whites
white sesames

Marinade

1 tsp salt
1 tsp ginger juice

Red vinegar syrup

3 tbsp maltose
3 tbsp red vinegar
2 tbsp Shaoxing wine
1 cup water

Preparations

1. Soak chicken wings in water until thawed. Drain and wipe dry. Add marinade and leave them for 2 hours. Blanch in boiling water. Drain and set aside.
2. Mix all ingredients for the red vinegar syrup. Cook in a hot water bath or double boiler until the maltose dissolves.
3. Put the chicken wings into the red vinegar syrup. Mix well. Arrange the chicken over a mesh wire rack. Leave them to air-dry for 4 hours. Brush egg white on them. Sprinkle with white sesames. Leave them to air-dry for 1 more hour.

Method

1. Heat enough oil in a wok for deep-frying. Check its temperature by inserting a wooden chopstick. The temperature is right if there are some bubbles coming out of the chopstick slowly. Put in the chicken wings. Deep-fry over low heat until golden and cooked through.

2. Turn the heat to high and fry the wings for 10 to 15 seconds before removing them from the oil. Drain well and serve.

Kitty's cooking tips

- When you deep-fry the chicken wings for the first time, be patient and fry them over low heat. Otherwise, they would turn too dark quickly while the inside is still raw.
- Before removing the chicken wings from oil and plating them, turn up the heat and fry them briefly. That would stop the wings from soaking up too much oil. They would be less greasy that way.

八爪魚乾有味煲仔飯

準備時間 1 小時

烹製時間 35 分鐘

材料

半肥瘦豬肉 2 兩
午餐肉 1 罐，細罐
南瓜 1/5 個
八爪魚乾 1/2 隻
冬菇 4 朵
紅葱頭 6 粒，切碎
紅米 2 兩
白米 6 兩

醃料

生抽 1 茶匙
生粉 1/2 茶匙
糖 1/2 茶匙
油少許

準備

1. 紅米用水浸 1 小時。
2. 半肥瘦豬肉切粒，與醃料拌勻，備用。
3. 南瓜切粒；八爪魚乾及冬菇分別用水浸軟，切粒。
4. 紅葱頭碎用油爆香，灑入鹽 1/4 茶匙，炒勻備用。

做法

1. 五餐肉切粗粒，鑊內放入油煎至金黃色備用。
2. 原鑊放入八爪魚乾及乾冬菇炒勻，放入豬肉粒同炒，炒至有香味，盛起。
3. 白米及紅米洗淨，放入瓦煲或電飯煲，加入適量水，排入南瓜同煲，滾起後，放入冬菇、豬肉及八爪魚，待水分收乾，加入午餐肉，再加入已炒香的紅葱頭，焗 10 分鐘即可。

三姐心得

用油炒香的紅葱頭，可儲存雪櫃。

- 午餐肉的油香，與冬菇及八爪魚乾的香味相配，配搭新穎，令整道煲仔飯帶有濃濃的鮮香、肉香味。
- 如用瓦煲，當聞到有飯焦香時，將瓦煲稍微傾側均勻地燒，並待白煙漸少時調至小火焗透。

Clay-pot rice with dried octopus

- Preparation time: 1 hour
- Cooking time: 35 minutes

Ingredients

75 g half-fatty pork
1 small can luncheon pork
1/5 pumpkin
1/2 dried octopus
4 dried shiitake mushrooms
6 shallots (finely chopped)
75 g red rice
225 g white rice
water

Marinade

1 tsp light soy sauce
1/2 tsp caltrop starch
1/2 tsp sugar
oil

Preparations

1. Soak red rice in water for 1 hour.
2. Dice the pork. Add marinade and mix well. Set aside.
3. Dice the pumpkin. Soak dried octopus and dried shiitake mushrooms in water separately until soft. Dice them.
4. Stir-fry the chopped shallots in some oil until fragrant. Season with 1/4 tsp of salt. Toss well. Set aside.

Method

1. Dice coarsely the luncheon pork. Fry in some oil until golden on both sides. Set aside.
2. In the same wok, stir-fry dried octopus and shiitake mushrooms together until well mixed. Put in the diced pork. Toss until fragrant. Set aside the mixture.
3. Rinse white rice and red rice. Put the mixture into a clay-pot or electric cooker. Add water. Arrange pumpkin on top. Bring to the boil. Add shiitake mushrooms, pork and octopus. Cook until the rice has absorbed all liquid. Arrange luncheon pork on top. Add stir-fried shallots. Turn off the heat. Cover the lid and leave it for 10 minutes. Serve.

Kitty's cooking tips

- The grease in the luncheon pork matches nicely with the flavours of shiitake mushrooms and dried octopus. This combination is creative and exciting, giving the whole pot of rice meaty flavours and briny-sweetness.
- If you're using a clay-pot, you may tilt the pot over the stove to heat its vertical sides when you begin to smell the smoky aromas of scorched rice.

準備時間
15
分鐘
烹製時間
25
分鐘
Braised fish head with chayote in oys
sauce
合掌瓜
蠔油燜魚雲

材料 

合掌瓜 2 個
魚雲 1 個
豆卜 6 個
薑 4 片

醃料 

生粉 2 湯匙
糖 1/2 茶匙
紹酒 1 湯匙

調味料 

紹酒 1 湯匙
生抽 1 湯匙
蠔油 1 湯匙
糖 1/2 茶匙

準備 

1. 合掌瓜刨皮，洗淨，去芯及切件。
2. 魚雲切開兩邊，放入醃料拌勻，備用。
3. 豆卜一開二，放入熱水略洗，備用。

做法 

1. 燒滾水，放入合掌瓜蒸 10 分鐘。
2. 燒熱油鍋，下薑片爆香；魚雲沾上生粉，放入鑊內煎香兩邊。
3. 放入紹酒、生抽及滾水 1 碗煮滾，排入蒸熟的合掌瓜及豆卜，加蓋煮 10 分鐘。
4. 最後，放入蠔油及糖調味，煮至汁液濃稠即可。

三姐心得

■ 豆卜用油炸起，下鍋煮前用熱水略沖洗，可減少吃入口的油膩感。

■ 合掌瓜先蒸 10 分鐘，能縮短燜煮之時間，好好控制魚雲及豆腐之烹調，令口感更佳。

Braised fish head with chayote in oyster sauce

■ Preparation time: 15 minutes
■ Cooking time: 25 minutes

Ingredients

2 chayotes
1 fish head
6 tofu puffs
4 slices ginger

Marinade

2 tbsp caltrop starch
1/2 tsp sugar
1 tbsp Shaoxing wine

Seasoning

1 tbsp Shaoxing wine
1 tbsp light soy sauce
1 tbsp oyster sauce
1/2 tsp sugar

Kitty's cooking tips

■ Tofu puffs are fried in oil. I prefer to cook them briefly in boiling water before using, so as to make them less greasy.
■ I steam the chayotes for 10 minutes first to shorten the braising time at last. That way, it's easier to control the cooking time so that the fish head and tofu puffs won't overcook.

Preparations

1. Peel the chayotes. Rinse well, core them and cut into chunks.
2. Cut the fish head into halves. Add marinade and mix well. Set aside.
3. Cut tofu puffs into halves. Blanch in hot water briefly. Drain well.

Method

1. Boil water in a steamer. Steam chayotes for 10 minutes.
2. Heat oil in a wok. Stir-fry ginger until fragrant. Coat fish head lightly in caltrop starch. Fry it until golden on both sides.
3. Add 1 tbsp of Shaoxing wine, 1 tbsp light soy sauce and 1 bowl of boiling water. Bring to the boil. Arrange the steamed chayotes and tofu puffs on top. Cover the lid and cook for 10 minutes.
4. Add 1 tbsp of oyster sauce and 1/2 tsp of sugar. Cook until the sauce thickens. Serve.

Steamed tofu egg white custard with assorted seafood

海皇蛋白豆腐

準備時間 10分鐘

烹製時間 10分鐘

材料

蛋白 5 個
濃無糖豆漿 1 杯，250 毫升
中蝦 3 隻
大帶子 2 粒
菜心或芥蘭 1 棵
蟹子 2 湯匙，裝飾
芫荽適量

蛋白豆腐調味料

鹽 1/2 茶匙
糖 1/2 茶匙

海鮮調味料

鹽 1/4 茶匙
胡椒粉少許
生粉 1 茶匙

準備

1. 蝦仁洗淨，去腸，用廚紙吸乾水分，切粒。帶子切碎。加鹽、胡椒粉和生粉略醃，加入約 1 湯匙蛋白拌勻。
2. 菜心或芥蘭切小粒。

做法

1. 蛋白放在大碗輕輕拂勻，倒入豆漿，用叉大力打至起泡，加入調味料拌勻。用篩過濾，蛋白豆漿倒入深碟內，用保鮮紙把碟子封好（小心別讓保鮮紙碰到豆漿）。
2. 蛋白豆漿以中火蒸 7 分鐘後收火，不要打開鑊蓋，原鑊再焗 2 分鐘。取出，就是蛋白豆腐底。
3. 起油鑊，以中火煎香蝦和帶子，放在蛋白豆腐上。
4. 菜粒下鑊炒熟，放在蝦和帶子周圍，放上蟹子和芫荽裝飾，即可上桌。

- 蒸蛋白豆腐時不要用太猛火，還要蓋上保鮮紙，以保持滑嫩。
- 夏天想有涼快的感覺，可將蛋白豆腐放入雪櫃冷藏 2 小時。
- 煎蝦和帶子時加入生粉和蛋白，作用有如脆漿，令海鮮香脆有汁。
- 如果想味道和色彩更豐富，可用粟米蓉 3 湯匙與已搓爛的熟蛋黃，煮成粟米汁淋在面。

Steamed tofu egg white custard with assorted seafood

- Preparation time: 10 minutes
- Cooking time: 10 minutes

Ingredients

5 egg whites
1 cup (250 ml) thick unsweetened soymilk
3 medium prawns (shelled)
2 large scallops
1 sprig Choy Sum or Chinese kale
2 tbsp sashimi-grade flying fish roe (as garnish)
coriander (as garnish)

Seasoning for custard

1/2 tsp salt
1/2 tsp sugar

Seasoning for seafood

1/4 tsp salt
ground white pepper
1 tsp caltrop starch

Preparations

1. Rinse the prawns and devein. Wipe dry with kitchen paper. Dice them and set aside. Dice the scallops. Add salt, ground white pepper and caltrop starch. Mix well and leave them for a while. Add 1 tbsp of egg white. Stir again.
2. Dice the Choy Sum or Chinese kale finely.

Method

1. Whisk the egg whites in a mixing bowl. Add soymilk. Whisk with a fork vigorously until foamy. Add seasoning. Mix well. Pass the mixture through a mesh sieve and save into a deep steaming dish. Cover with microwave-safe cling film. Use care not to let the cling film touch the soymilk and egg white mixture.
2. Steam the egg white and soymilk over medium heat for 7 minutes. Turn off the heat and leave it in the steamer for 2 minutes without opening the lid. Remove the custard from the steamer.
3. Heat oil in a wok and fry the prawns and scallops over medium heat until cooked through. Transfer them over the steamed custard from step 2.
4. In the same wok, stir-fry diced Choy Sum or Chinese kale until cooked through. Arrange over the custard around the prawns and scallops. Garnish with flying fish roe and coriander. Serve.

Kitty's cooking tips

- Do not steam the custard over high heat. Make sure you cover the dish with cling film to keep the custard silky and avoid overcooking.
- To serve this dish as a cold dish for summer, refrigerate the custard for 2 hours before putting the prawns and scallops over.
- When I marinate the prawns and scallops, I add caltrop starch and egg white. They will form a coating around the prawns and scallops like a batter, sealing in the juices and keeping them moist.
- For extra layering of texture and a splash of colour, mash the yolk of a hard-boiled egg. Add 3 tbsp of cream style sweet corn. Heat it up and dribble this sauce all over before serving.

Spicy prawns with fermented beancurd sauce

香辣腐乳蝦

烹製時間
10
分鐘

材料

中蝦 8 隻
芹菜 1 棵
紅椒 1 隻
蒜蓉 1 湯匙
芫荽 1 棵

調味料

辣椒腐乳 3 件
糖 1 茶匙

準備

1 中蝦洗淨，吸乾水分，剪去鬚腳及眼，剒開蝦背，起背腸。

2 腐乳壓爛，加糖搓成醬。

3 芹菜切長粒；紅椒切絲，備用。

做法

1 起油鑊，放入蝦走油或半煎炸至半熟。

2 倒去多餘的油，燒熱鑊底油，爆香蒜蓉和腐乳醬，下蝦翻炒，最後下芹菜粒和紅椒絲炒勻，上碟，放上芫荽裝飾。

三姐心得

- 這道菜宜選用肉質較爽的冰鮮南美蝦。
- 炸蝦必須剪去眼，以免油炸時爆開產生危險。

Spicy prawns with fermented beancurd sauce

- Preparation time: 15 minutes
- Cooking time: 10 minutes

Ingredients

8 medium prawns
1 sprig Chinese celery
1 red chilli
1 tbsp grated garlic
1 sprig coriander

Seasoning

3 cubes fermented beancurd with chilli
1 tsp sugar

Preparations

1. Rinse the prawns and wipe dry. Trim off the antennae, legs and eyes. Cut along the back and devein.
2. Mash the fermented tarocurd. Add sugar and mix well.
3. Coarsely dice the Chinese celery. Set aside. Finely shred chilli.

Method

1. Heat oil in a wok. Blanch the prawns in oil or semi-deep fry them until half cooked.
2. Drain most of the oil. Heat the wok again. Stir-fry garlic and fermented beancurd until fragrant. Put the prawns back in and toss well. Sprinkle with Chinese celery and red chilli. Toss again. Garnish with coriander. Serve.

Kitty's cooking tips

- For this recipe, I prefer using chilled white-leg prawns from South America, for their springy texture.
- Make sure you trim off the eyes of the prawns before deep-frying them. Otherwise, their eyes may burst and the oil may splatter.

Deep-fried abalones in peppered salt

脆皮鮑魚

材料

鮮鮑魚 6 隻
紅椒 1 隻，切絲
炸蒜 1 湯匙

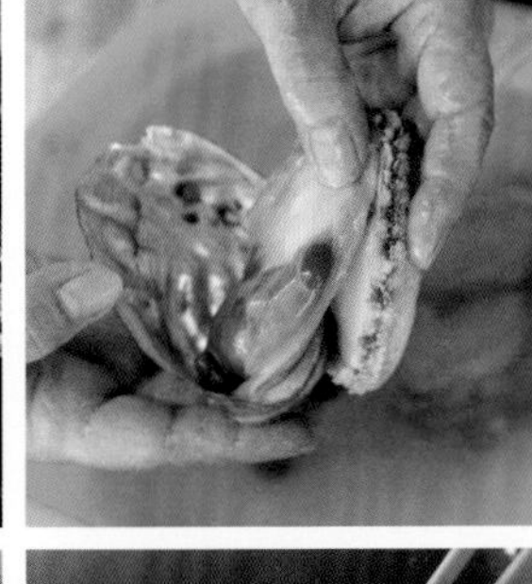

脆漿料

蛋白 1 個
炸粉 1 湯匙
油 1 湯匙

調味料

椒鹽 1/2 茶匙

準備

1. 鮑魚洗淨，用熱水略灼，起走鮑魚腸和殼，洗淨抹乾，裹上少許生粉。
2. 蛋白和炸粉調勻，加油攪勻，放置半小時。

做法

1. 鮑魚蘸上脆漿，放入燒至八分熱的油內，以大火炸約 2 分鐘，見金黃色即撈起，瀝去油分。
2. 鮑魚一切開二，排放碟上，灑上椒鹽調味，再放上炸蒜和紅椒絲即成。

三姐心得

- 鮑魚勿炸得過熟，八成熟即可，切開可見溏心為佳。
- 炸鮑魚外脆內軟兼有汁，不用蘸任何醬汁已很美味。

Deep-fried abalones in peppered salt

- Preparation time: 10 minutes
- Cooking time: 5 minutes

Ingredients

6 live abalones
1 red chilli (shredded)
1 tbsp deep-fried garlic bits

Deep-frying batter

1 egg white
1 tbsp deep-frying batter mix
1 tbsp oil

Seasoning

1/2 tsp peppered salt

Preparations

1. Rinse the abalones. Blanch in boiling water briefly. Shell them and remove the innards. Rinse and wipe dry. Coat them lightly in caltrop starch.
2. To make the deep-frying batter, mix egg white with deep-frying batter mix first. Then add oil and stir well. Leave it to stand for 30 minutes.

Method

1. Dunk the abalones into the deep-frying batter. Heat enough oil in a wok for deep-frying up to 200°C over high heat. Put in the abalones one by one and deep-fry for about 2 minutes until golden. Drain.
2. Slice each abalone in half. Arrange on a serving plate. Sprinkle with peppered salt. Garnish with deep-fried garlic bits and red chilli. Serve.

Kitty's cooking tips

- Do not overcook the abalone. It should be slightly undercooked to stay juicy and tender. When you slice it, you should see a very small core that is undercooked.
- The abalones are crunchy on the outside and juicy on the inside. You don't need any dipping sauce and they taste awesome as they are.

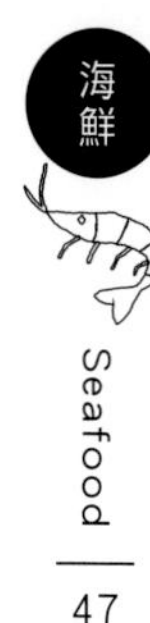

Stir-fried mung bean vermicelli with clams

鮮蜆肉炒馬尾粉絲

材料

馬尾粉絲 120 克
蜆肉 150 克
櫻花蝦 2 湯匙
鹹蛋黃 1 個
銀芽 40 克
芹菜 1 棵
芫荽 1 棵
紅椒 1/2 隻
蒜頭 1 瓣

調味料

蠔油 1 茶匙
XO 醬 1 茶匙
生抽 1 茶匙
老抽 1/2 茶匙
糖少許
水適量

準備

1. 粉絲用冷水浸軟。
2. 蜆肉洗淨，吸乾水分。
3. 芹菜、芫荽洗淨摘葉，切長粒；紅椒切粒；蒜頭剁茸。
4. 鹹蛋蒸熟，取黃壓碎。

做法

1. 燒熱油鑊，下油爆香蒜蓉，放下蜆肉炒香，盛出。
2. 粉絲下鑊炒軟，取出備用。
3. 再起油鑊，以慢火炒香櫻花蝦；轉回中火，加入鹹蛋黃、XO 醬及其他調味料炒勻，加入粉絲，邊炒邊加入少量水，至粉絲不再成團，顏色均勻。
4. 放下銀芽炒勻，再下蜆肉、芹菜、芫荽、紅椒炒勻即可上碟。

三姐心得

- 炒粉絲時不斷用筷子挑鬆就不會黐鑊。粉絲黐作一團表示未吸夠水分，要邊炒邊加水，至粉絲鬆散即成。
- 馬尾粉絲（見圖右）是雜貨舖發售的散裝粉絲，是高級食肆常用的食材；這種粉絲質地幼滑爽韌，浸後不會發脹變爛。也可用一般的龍口粉絲（見圖左）或其他粉絲替代。

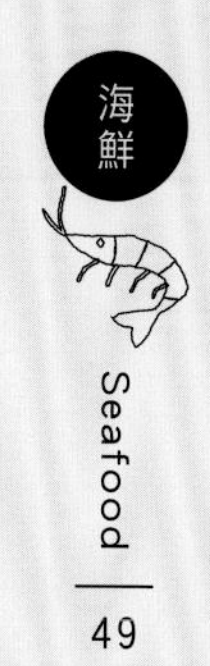

Stir-fried mung bean vermicelli with clams

- Preparation time: 20 minutes
- Cooking time: 10 minutes

Ingredients

120 g "horse-tail" mung bean vermicelli
150 g shelled clams
2 tbsp Sakura shrimps
1 salted egg yolk
40 g mung bean sprouts
1 sprig Chinese celery
1 sprig coriander
1/2 red chilli
1 clove garlic

Seasoning

1 tsp oyster sauce
1 tsp XO sauce
1 tsp light soy sauce
1/2 tsp dark soy sauce
sugar
water

Preparations

1. Soak mung bean vermicelli in cold water until soft. Drain and set aside.
2. Rinse the clams. Wipe dry.
3. Rinse Chinese celery and coriander. Pick off and discard the leaves. Dice the stems coarsely. Set aside. Dice the chilli. Finely chop the garlic.
4. Steam a salted egg till done. Shell and separate the yolk from the white. Mash the yolk and set aside.

Method

1. Heat a wok and add oil. Stir-fry garlic until fragrant. Stir-fry the clams until fragrant. Set aside.
2. In the same wok, stir-fry mung bean vermicelli until soft. Set aside.
3. Heat the wok again and add oil. Stir-fry the Sakura shrimps over low heat until fragrant. Turn the heat to medium. Add salted egg yolk and seasoning. Toss well. Put the mung bean vermicelli back in. Keep tossing while adding a little water at a time until the mung bean vermicelli separate from one another and pick up the colour of the seasoning evenly.
4. Add mung bean sprouts and toss again. Sprinkle with clams, Chinese celery, coriander and red chilli. Toss to mix evenly. Serve.

Kitty's cooking tips

- Instead of a spatula, use a pair of chopsticks to toss the mung bean vermicelli. They are less like to break and stick to the wok this way. If they clump together in one mass, it means they are too dry and you need to add more water while tossing them.
- "Horse-tail" mung bean vermicelli is sold in bulk in traditional grocery stores and is commonly used in upmarket eateries. They are silky but resilient in texture. They won't swell too much and turn mushy even after soaked in water. If you can't get "horse-tail" mung bean vermicelli, you can replace it with regular pre-packaged ones from Longkou or any other varieties.

Braised crab with melon and glutinous rice cake

白瓜年糕燜蟹

準備時間 20 分鐘

烹製時間 20 分鐘

材料 

蟹 1 隻
白瓜 1 條
上海年糕 150 克
芹菜 1 棵
草菇 10 粒
熱上湯 1/2 碗
紅椒絲少許
薑 3 塊
蒜頭 3 瓣

調味料 

蠔油 1 湯匙
糖 1/2 茶匙
鹽 1 茶匙

準備 

1. 年糕切成約手指粗幼的長條，用熱水浸軟備用，下鑊前撈起瀝乾。
2. 白瓜可去皮或不去皮，去核，斜切成厚件。
3. 芹菜切段；鮮草菇出水，對半切開。
4. 蟹劏好洗淨，起蓋，斬成 6 件。

做法

1. 起油鑊，先下薑片爆香，加入蒜頭爆香。在蟹膏和蟹件拍上生粉，下鑊炸至半熟。
2. 加入白瓜煮片刻，下草菇和年糕，倒入熱上湯，下調味料，加蓋焗 2 至 3 分鐘。
3. 加入芹菜（喜歡辣的，可加辣椒），撒入胡椒粉。
4. 上碟時，先夾起蟹件，不收火，讓汁收乾些，將配菜放碟中，再將蟹排在上即可。

三姐心得

■ 白瓜是一種廣東常見的夏季時令食材，味較淡，蟹的鮮味既可將白瓜的清甜「調」出來，白瓜和年糕又完全吸收蟹汁的美味。

Braised crab with melon and glutinous rice cake

- Preparation time: 20 minutes
- Cooking time: 20 minutes

Ingredients

1 crab
1 oriental pickling melon
150 g Shanghainese glutinous rice cake
1 sprig Chinese celery
10 straw mushrooms
1/2 bowl stock (heated)
shredded red chilli
3 slices ginger
3 cloves garlic

Seasoning

1 tbsp oyster sauce
1/2 tsp sugar
1 tsp salt

Preparations

1. Cut the glutinous rice cake into short strips about the thickness of a finger. Soak in hot water. Drain right before using.
2. You may choose to keep the skin of the oriental pickling melon or peel it. De-seed and slice diagonally into thick pieces.
3. Cut Chinese celery into short lengths. Set aside. Blanch straw mushrooms in boiling water briefly. Cut them in half.
4. Dress the crab and rinse well. Separate the carapace from the body. Chop the body into 6 pieces.

Method

1. Heat oil in a wok and stir-fry the ginger until fragrant. Add garlic and fry till fragrant. Coat the crab carapace and pieces in caltrop starch lightly. Deep-fry in oil until half-cooked.
2. Add the oriental pickling melon. Cook briefly. Put in straw mushrooms and glutinous rice cake. Pour in the hot stock. Add seasoning. Cover the lid and cook for 2 to 3 minutes.
3. Add Chinese celery and red chilli (optional). Sprinkle with ground white pepper.
4. Set aside the crab first. Keep cooking the glutinous rice cake and melon to reduce the sauce further. Then transfer them onto a serving plate. Arrange the crab on top. Serve.

Kitty's cooking tips

■ Oriental pickling melon is a commonly seen ingredient in Guangdong region during summer. It is blander in flavour than cucumber. The seafood richness of crab can bring out the sweetness in oriental pickling melon. Both the melon and the rice cake would slurp up the juices of the crab and become flavourful.

Scrambled egg with fresh yam and scallops

鮮淮山帶子炒蛋

材料 

雞蛋 4 個
鮮淮山小半枝
大帶子 3 粒
蝦仁 80 克
生粉 1/2 茶匙

準備 

1. 蝦仁切成小粒，用鹽 1/2 茶匙拌匀，醃 3 小時。
2. 雞蛋打入大湯碗內，不要拂匀，加入鹽、油，用筷子戳破蛋黃，輕輕攪一下。
3. 帶子洗淨，吸乾水分，切粗粒，用生粉拌匀。
4. 鮮淮山去皮，切絲，汆水半分鐘，用清水沖洗至冷。

做法 

1. 燒熱油鑊，放入鹹蝦粒和帶子，煎至金黃色；加入淮山絲煎香，取出。
2. 燒熱鑊，加入約 3 湯匙油，倒入雞蛋和已煎香的材料，快手炒至雞蛋半熟，即可上碟。

- 鮮淮山切開後很易變黃變黑，若不是馬上下鍋，要汆水或用清水浸泡。
- 雞蛋不拂匀，炒起來味道較香濃。

Scrambled egg with fresh yam and scallops

■ Preparation time: 10 minutes
■ Cooking time: 5 minutes

Ingredients

4 eggs
1/2 stem fresh yam
3 large scallops
80 g shelled shrimps
1/2 tsp caltrop starch

Preparations

1. Dice the shrimps. Add 1/2 tsp of salt and mix well. Leave them for 3 hours.
2. Crack the eggs into a mixing bowl. Add salt and oil. Poke the egg yolks with chopsticks. Roughly stir them (without completely incorporating the yolks with the whites).
3. Rinse the scallops. Wipe dry and dice coarsely. Add caltrop starch and mix well.
4. Peel the yam and finely shred it. Blanch in boiling water for 30 seconds. Rinse in cold water until cool to the touch.

Method

1. Heat a wok and add oil. Fry shrimps and scallops until golden. Add yam and toss until lightly browned. Set aside.
2. Heat the wok again and add 3 tbsp of oil. Pour in the eggs and the stir-fried ingredients from step 1. Toss quickly until the eggs are half-set. Serve.

Kitty's cooking tips

■ Fresh yam turns brown easily if left in the air after peeled and sliced. If you don't intend to cook it right away, keep it in water or blanch it first.
■ You don't whisk the eggs too well. When the yolks and whites are still separated, the eggy flavour tends to be stronger after cooked.

Sweet and sour prawns in liquorice plum sauce

話梅咕嚕蝦

準備時間 20 分鐘

烹製時間 10 分鐘

材料

大蝦 10 隻
青紅椒、洋葱各 1/2 個

話梅汁料 

牛油 2 茶匙
咕嚕汁 1/2 碗（做法見「心得」）
話梅粉 1 茶匙
蜜糖 1 茶匙

蝦醃料 

糖 1/2 茶匙
鹽 1/2 茶匙
胡椒粉少許

炸漿 

雞蛋 1 個
生粉 1 湯匙

準備 

1. 大蝦洗淨，吸乾水分，去背腸。加入醃料拌勻，醃 15 分鐘。
2. 青紅椒切角；洋葱切塊，備用。
3. 雞蛋拂勻，加入生粉 1 湯匙，拌勻成炸漿。

做法

1. 起油鑊，放入青紅椒和洋葱走油，撈起備用。
2. 蝦放入蛋漿撈勻，再放入熱油中炸至八成熟。
3. 另起鑊，放下牛油煮溶，加入其他話梅汁材料拌勻，下青紅椒、洋葱及蝦，快手翻炒，即可上碟。

三姐心得

- 話梅粉可以用話梅切碎代替。
- 自製咕嚕汁：茄汁 1 碗，白醋 1 碗，山楂餅 10 片，OK 汁 2 湯匙，喼汁 2 湯匙，黃糖 2 湯匙，如有茄膏可放 1 湯匙，攪勻，煮滾即可。

Sweet and sour prawns in liquorice plum sauce

- Preparation time: 20 minutes
- Cooking time: 10 minutes

Ingredients

10 large prawns
1/2 green bell pepper
1/2 red bell pepper
1/2 onion

Liquorice plum sauce

2 tsp butter
1/2 bowl home-made sweet and sour sauce (recipe included in tips)
1 tsp liquorice plum powder
1 tsp honey

Marinade

1/2 tsp sugar
1/2 tsp salt
ground white pepper

Deep-frying batter

1 egg
1 tbsp caltrop starch

Preparations

1. Rinse the prawns and wipe dry. Devein. Add marinade and mix well. Leave them for 15 minutes.
2. Cut green and red bell pepper into wedges. Cut onion into chunks.
3. Whisk the egg. Add 1 tbsp of caltrop starch and mix well. This is the deep-frying batter.

Method

1. Heat oil in a wok. Fry the bell peppers and onion briefly. Set aside.
2. Dunk the prawns into the deep-frying batter. Deep-fry in hot oil until medium-well done.
3. In another wok, melt the butter. Put in all sauce ingredients and mix well. Put in bell peppers, onion and prawns. Toss quickly to coat evenly. Serve.

Kitty's cooking tips

- You may use de-seeded and shredded dried liquorice plums instead of the plum powder.
- To make your own sweet and sour sauce, mix together 1 bowl ketchup, 1 bowl rice vinegar, 10 haw flakes, 2 tbsp OK steak sauce, 2 tbsp Worcestershire sauce, 2 tbsp light brown sugar, and 1 tbsp tomato paste (optional). Bring to the boil in a saucepan and mix well.

Seafood casserole with Napa cabbage and wontons

津白雲吞海鮮鍋

材料 

黃芽白 1/2 棵
菜肉雲吞 6 粒
豆卜 6 個
花蟹 1 隻
花蛤 300 克
薑 3 片
芹菜 2 棵
上湯或清水 1 1/2 碗

調味料 

胡椒粉適量

準備 

1. 菜肉雲吞灼熟；豆卜剪開反轉，用油炸脆，備用。
2. 黃芽白洗淨，切件；芹菜切粒。
3. 蟹起蓋，去鰓和胃，洗淨，斬件。

做法

1. 在大瓦煲煮滾上湯，放入薑片和黃芽白煮軟。
2. 放上蟹件，四周放上豆卜和雲吞，煮滾，最後放上花蛤，加蓋煮 3 分鐘。
3. 關火，放入芹菜粒，加蓋焗片刻；撒上胡椒粉調味，原煲上桌。如喜歡辣，可加入指天椒。

- 花蛤用膠袋連海水盛好，放雪櫃 3 至 4 小時，自會張開，洗淨沙粒就很乾淨。

Seafood casserole with Napa cabbage and wontons

- Preparation time: 20 minutes
- Cooking time: 20 minutes

Ingredients

1/2 long Napa cabbage
6 wontons with cabbage and pork filling
6 tofu puffs
1 swimmer crab
300 g live clams
3 slices ginger
2 sprigs Chinese celery
1 1/2 bowl stock (or water)

Seasoning

ground white pepper

Preparations

1. Blanch the wontons in boiling water till done. Drain and set aside. Cut the tofu puffs in half. Flip them inside out. Deep-fry in oil until crispy. Drain and set aside.
2. Rinse the Napa cabbage. Cut into pieces. Set aside. Dice the Chinese celery.
3. Dress the crab and separate the carapace from the body. Remove the gills and the stomach. Rinse and chop the body into pieces.

Method

1. Boil the stock in a casserole pot. Add ginger and Napa cabbage. Cook until soft.
2. Put in the crab. Arrange tofu puffs and wontons around. Bring to the boil and add clams. Cover the lid and cook for 3 minutes.
3. Turn off the heat and add Chinese celery. Cover the lid and leave it briefly. Sprinkle with ground white pepper. Serve the whole pot. Optionally, add a sliced bird's eye chilli for an extra kick.

Kitty's cooking tips

- To make the clams spit out the sand, just keep them in a plastic bag with sea water (or salted water). Leave them in the fridge for 3 to 4 hours. The clams would open their shells and pump water in and out of their body for breathing. The sand will be removed in due course.

Razor clams and pork tripe in black bean sauce over fried rice noodles

豉椒豬肚炒蟶子煎脆米

準備時間 40 分鐘

烹製時間 10 分鐘

海鮮 Seafood

材料

大蟶子 4 隻
急凍豬肚 1/4 個
青、紅椒各 1/2 個
紅葱頭 2 粒
金菇 1/2 包
蒜粒 4 瓣
豆豉 1 湯匙
米粉 1/2 個，用暖水浸至軟身

調味料

生抽 1 湯匙
老抽少許
鹽 1/4 茶匙
糖 1/2 茶匙
麻油少許
紹酒 1/2 湯匙
生粉水

三姐心得

■ 豬肚要爽而不韌，煮時切勿過火，煮至能用筷子刺入即可。如要較腍可放在熱水焗一會，加鹽和胡椒粉調味；如要爽口則取出浸在有鹽的冰水中。

準備 

1. 在平底鑊燒熱 2 湯匙油，將米粉逐少鋪放鑊內，鋪成薄薄的一層，以中火煎至米粉底面金黃香脆，取出，反轉放在架上待涼。
2. 蟶子洗淨，去腸，放入熱水中 10 秒，取出；馬上放入冰水起肉並再次清洗，瀝乾水分。蟶子殼放入熱水灼 1 至 2 分鐘，撈起排放碟上。
3. 急凍豬肚洗淨，放入滾水煮半小時，用筷子插入，能插穿即可，撈出放涼，切幼條。
4. 紅葱頭去衣、切片；青、紅椒切塊。蒜頭去衣、拍碎，與 1/2 湯匙豆豉一起用刀柄舂爛。

做法

1. 起油鑊，爆香紅葱頭，下青、紅椒略炒，盛出。
2. 再下油，爆豬肚至金黃色及有香味，推到鑊邊。加油，下蒜蓉豆豉爆香，加入原粒豆豉和金菇煸炒至軟，下蟶子及調味料快手翻炒，灒酒，夾出蟶子以免過火變韌，下青、紅椒及少許鹽炒勻，最後將蟶子回鑊拌勻，即可上碟。
3. 將已涼的脆米粉分成 4 至 5 塊，放在碟邊伴吃。

Razor clams and pork tripe in black bean sauce over fried rice noodles

- Preparation time: 40 minutes
- Cooking time: 10 minutes

Ingredients

4 large razor clams
1/4 frozen pork tripe
1/2 green bell pepper
1/2 red bell pepper
2 shallots
1/2 pack enokitake mushrooms
4 cloves garlic
1 tbsp fermented black beans
1/2 bundle rice vermicelli (soaked in warm water till soft)

Seasoning

1 tbsp light soy sauce
dark soy sauce
1/4 tsp salt
1/2 tsp sugar
sesame oil
1/2 tbsp Shaoxing wine
caltrop starch slurry (1 part caltrop starch mixed with 1 part water)

Preparations

1. Heat 2 tbsp of oil in a pan. Arrange the rice vermicelli evenly on the pan to form an even layer. Fry over medium heat until golden and crispy on one side. Remove and leave it to cool upside down on a wire rack.
2. Rinse the razor clams and remove the innards. Blanch in boiling water for 10 seconds. Dunk into ice water and shell them. Rinse the flesh in cold water. Drain well and set aside. Blanch the shells in boiling water for 1 to 2 minutes. Drain well and arrange on a serving plate.
3. Rinse the pork tripe. Blanch in boiling water for 30 minutes. Insert a chopstick into the thickest part. It is done if the chopstick can poke through easily. Drain and let cool. Cut into thin strips.
4. Peel and slice shallots. Cut all bell peppers into strips. Peel and crush the garlic. Put garlic into a bowl. Add 1/2 tbsp of fermented black beans. Crush with the handle of a cleaver (or with mortar and pestle).

Method

1. Heat oil in a wok and fry the shallots until fragrant. Put in the bell peppers and toss briefly. Set aside.
2. In the same wok, heat some oil and fry the pork tripe until golden and fragrant. Push to one side of the wok. Add oil and stir-fry garlic and black bean mixture until fragrant. Add the remaining fermented black beans and enokitake mushrooms. Stir over high heat till soft. Put in the shelled razor clams and seasoning. Toss quickly to mix well. Sprinkle with wine. Remove the razor clams to stop them from being overcooked. Put in the bell peppers and a pinch of salt. Toss again. Put the razor clams back in. Toss and transfer the mixture over the clam shells.
3. Cut the fried rice vermicelli into 4 or 5 pieces. Arrange on the side. Serve.

Kitty's cooking tips

■ Do not overcook the pork tripe. It should retain a lovely chew yet without being too chewy or rubber. When you can easily poke it through with a chopstick, it is cooked through. If you want it to be a bit softer, turn off the heat and leave it in the hot water for a while longer with the lid closed. Season with salt and ground white pepper. If you want it to be springier in texture, soak it in ice water with a pinch of salt.

Scrambled egg with dried scallop, crabmeat and bean sprouts

炒桂花瑤柱

準備時間
10
分鐘

烹製時間
15
分鐘

材料 

雞蛋 3 個
雞蛋黃 1 個
乾瑤柱 80 克
蟹肉 60 克
銀芽 40 克
韭菜花少許
紅椒絲少許

調味料

鹽 1 茶匙
糖 1/2 茶匙
胡椒粉少許

準備 

1. 瑤柱洗淨、浸軟，加少許水蒸脸，撕成幼絲，可炸香。
2. 雞蛋及蛋黃打入大碗內，加少許鹽和 1 茶匙油拂勻。

做法 

1. 起油鑊，將銀芽和韭菜花過油，取出。下蟹肉煎香，加入少許鹽調味，盛起備用。
2. 另起油鑊，調至中火，下雞蛋，先用筷子不停攪動炒蛋至熟，收慢火，改用鑊鏟不停推動搓壓，將蛋弄成細小的碎粒。
3. 調至中火，加入蟹肉、瑤柱絲、銀芽和韭菜花炒勻，最後下紅椒絲，轉大火，翻炒幾下，即可上碟。

三姐示範

■ 桂花是指雞蛋細細粒，似桂花形狀，炒蛋時需要用鑊鏟不斷推壓雞蛋成小塊狀。

Scrambled egg with dried scallop, crabmeat and bean sprouts

- Preparation time: 10 minutes
- Cooking time: 15 minutes

Ingredients

3 eggs
1 egg yolk
80 g dried scallops
60 g crabmeat
40 g mung bean vermicelli
flowering chives
shredded red chilli

Seasoning

1 tsp salt
1/2 tsp sugar
ground white pepper

Preparations

1. Rinse and soak the dried scallops in water until soft. Transfer into a steaming bowl and add a little water. Steam until soft. Tear into fine shreds. Optionally, you may deep-fry them till crispy.
2. Put eggs and egg yolk into a mixing bowl. Add a pinch of salt and 1 tsp of oil. Whisk well.

Method

1. Heat a wok and add oil. Blanch the mung bean sprouts and flowering chives in oil briefly. Set aside and drain excess oil. Fry the crabmeat until fragrant. Season with a pinch of salt. Set aside.
2. Heat another wok and add oil. Turn to medium heat. Pour in the egg mixture and stir with chopsticks continuously to make egg strands. Turn to low heat. Use a spatula to toss and egg strands and break them into small bits.
3. Turn to medium heat. Put in the crabmeat, dried scallops, mung bean sprouts and flowering chives. Toss to mix well. Add red chilli at last. Turn to high heat. Toss a few times. Serve.

Kitty's cooking tips

- As opposed to the usual scrambled egg that is half-set and silky smooth, the egg in this dish is supposed to be dry and fully cooked. The egg is broken down into small bits with a spatula so that they look like osmanthus flowers.

Deep-fried squid and veggie rings in peppered salt

椒鹽三式圈

準備時間
10
分鐘

烹製時間
3
分鐘

材料

鮮魷 1 隻
涼瓜 1 個，長型
紫洋葱 1 個

脆漿材料

日式炸粉 1/2 碗
油 2 至 3 湯匙
水半碗
雞蛋黃 1 個

調味料

椒鹽適量

準備

1. 炸粉加油慢慢拌勻，加入少量水，攪至沒有粉粒，加入蛋黃再拌，慢慢加水，將粉漿調至流質狀備用。
2. 魷魚洗淨，瀝乾水分，分開頭尾，魷魚身切圈。
3. 涼瓜切圈、去瓤；洋葱切圈。

做法

1. 涼瓜圈蘸薄漿粉；洋葱先用生粉抓一下，再蘸薄漿粉。待油燒至六成熱（約 150℃），放入涼瓜圈炸約 15 秒，加入洋葱圈，至涼瓜和洋葱呈金黃色，撈出瀝油。
2. 保持油溫六成熱，魷魚圈先用生粉抓一下，再蘸薄漿粉，下鑊炸約 45 秒至金黃色，撈起瀝油。
3. 撒上椒鹽調味，排上碟享用。

三姐心得

- 炸漿用手抓攪較易達到均勻及幼滑的效果。
- 可隨意選用其他蔬菜炸脆，如燈籠椒、四季豆等。
- 先炸菜類，後肉類，油會較為乾淨。

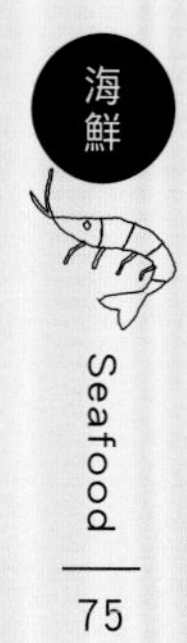

Deep-fried squid and veggie rings in peppered salt

- Preparation time: 10 minutes
- Cooking time: 3 minutes

Ingredients

1 squid
1 bitter melon (elongated-shaped)
1 purple onion

Deep-frying batter

1/2 bowl Japanese deep-frying flour mix
2 to 3 tbsp oil
1/2 water
1 egg yolk

Seasoning

peppered salt

Kitty's cooking tips

- I prefer mixing the deep-frying batter with my hand. You can make sure it's well mixed and lump free that way.
- Instead of bitter melon and onion, you may also use other veggies instead, such as bell pepper rings or French beans.
- I always deep-fry the veggies first before deep-frying proteins. The oil tends to be clearer with fewer dark burnt bits that way.

Preparations

1. Put the deep-frying flour mix into a bowl. Slowly add oil and keep stirring. Add a little water and stir until lump-free. Add egg yolk and stir again. Slowly add water to thin it out into a runny consistency.
2. Rinse the squid and rinse well. Pull the head and tentacles out. Then slice the body into rings.
3. Cut the bitter melon into rings and remove the seeds. Cut onion into rings.

Method

1. Dip the bitter melon rings into the batter. Mix well. Coat onion rings lightly in caltrop starch. Then dip them into the batter. Heat enough oil for deep-frying up to 150°C. Put in the bitter melon rings one by one. Fry for 15 seconds. Put in the onion rings one by one. Fry them till both the bitter melon and onion rings are golden. Remove them from the oil with a strainer ladle. Drain excess oil.
2. In the same wok, heat the oil up to 150°C again. Coat the squid rings lightly in caltrop starch. Dip them into the batter. Deep-fry for about 45 seconds until golden. Drain.
3. Sprinkle with peppered salt. Arrange on a serving plate. Serve.

Steamed seafood and salted pork bones
in winter melon boat
鹹骨海鮮
冬瓜船
準備時間
20
分鐘
烹製時間
45
分鐘

材料

袖珍冬瓜 1 個
排骨 150 克
蟹 1 隻，小
中蝦 4 隻
蜆適量
鮮蓮子 15 至 20 粒
乾雜豆 2 湯匙
夜香花適量

準備

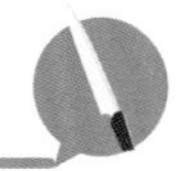

1. 冬瓜在 1/3 處橫切開，去瓤及挖去部分瓜肉，造成船型。
2. 排骨用鹽 1 茶匙醃 4 至 6 小時。
3. 乾雜豆洗淨，浸 4 小時，瀝乾。
4. 蓮子用牙籤去芯，浸水 1 小時。

做法

1. 鹹排骨、雜豆和蓮子放入冬瓜內，以荷葉墊住蒸 35 分鐘。將蝦、蟹放在面，再蒸 7 至 8 分鐘，最後放入蜆蒸 2 分鐘。
2. 夜香花放入滾水灼一下，撈出，灑在冬瓜船上，享用。

三姐心得

- 新鮮蓮子夏天才上市，蒸熟來吃最原汁原味，質地粉糯，味道清甜。找不到新鮮蓮子，可用乾白蓮代替。份量是鮮蓮子的一半，預浸 4 至 6 小時。
- 夜香花必須灼一會，才能散發清幽的香氣。

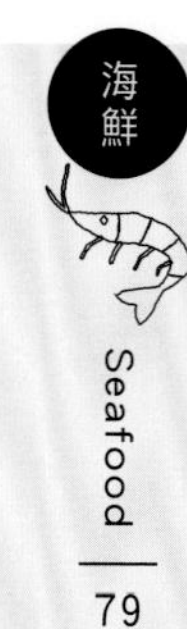

Steamed seafood and salted pork bones in winter melon boat

- Preparation time: 20 minutes
- Cooking time: 45 minutes

Ingredients

1 mini winter melon
150 g pork ribs
1 small crab
4 medium shrimps
live clams
15 to 20 fresh lotus seeds
2 tbsp assorted dried beans
Tonkin jasmine buds

Preparations

1. Put the winter melon on a counter the long side down. Make a horizontal cut along the top third of the melon. Scoop out and discard the seeds. Scoop out part of the flesh. Shape into a boat.
2. Rub 1 tsp of salt evenly on the pork ribs. Leave them for 4 to 6 hours. This is the salted pork ribs.
3. Soak the assorted dried beans in water for 4 hours. Drain well.
4. Remove the core of lotus seeds with toothpick. Soak in water for 1 hour.

Method

1. Put the salted pork ribs, assorted dried beans and lotus seeds into the winter melon. Cover with a lotus leaf and steam for 35 minutes. Arrange the shrimps and crab on top. Steam for 7 to 8 minutes. Lastly put in the clams. Steam for 2 minutes.
2. Blanch the Tonkin jasmine buds in boiling water briefly. Drain and sprinkle over the winter melon boat. Serve.

Kitty's cooking tips

- Fresh lotus seeds are only available in summer. They are best steamed as it allows their natural flavours to come through. If you can't get fresh lotus seeds, use dried white lotus seeds instead. You only need half the weight if you use dried ones. Soak them for 4 to 6 hours before using.
- You must blanch the Tonkin jasmine buds in hot water briefly for them to release their fragrance.

Fried prawns and shredded roast duck in plum sauce

火鴨絲梅子蝦

準備時間 10 分鐘

烹製時間 10 分鐘

材料

中蝦 10 隻
火鴨 150 克
芹菜粒 1 湯匙

蝦醃料

鹽少許
胡椒粉少許
生粉 1/2 茶匙

梅子醬材料

梅子 2 粒
鴨醬 2 湯匙
片糖粉 1 茶匙

準備

1. 中蝦洗淨，吸乾水分，剪去鬚腳及眼，劏開蝦背，起背腸。
2. 火鴨切絲；梅子用叉壓爛。

做法

1. 起油鑊，將蝦煎好，取出。
2. 下 1 湯匙油燒熱，放下火鴨絲慢火爆香，下梅子、鴨醬和片糖粉煮滾，將蝦回鑊翻炒一下，下芹菜粒拌勻，即可上碟。或用青瓜片、芫荽、紅椒絲等伴碟。

- 煎蝦和火鴨都是惹味的食材，搭配在一起，味道融和且互相提味。
- 吃不完的火鴨可烹調此餸，以免浪費食材。

Fried prawns and shredded roast duck in plum sauce

- Preparation time: 10 minutes
- Cooking time: 10 minutes

Ingredients

10 medium prawns
150 g roast duck
1 tbsp diced Chinese celery

Marinade

salt
ground white pepper
1/2 tsp caltrop starch

Plum sauce

2 pickled plums in brine
2 tbsp plum sauce for roast duck
1 tsp raw cane sugar

Preparations

1. Rinse the prawns and wipe dry. Trim off the antennae, legs and eyes. Cut along the back and devein.
2. De-bone the duck and shred the flesh. Drain and de-seed the pickled plums. Mash with a fork.

Method

1. Heat oil in a wok and fry the prawns until done. Set aside.
2. Heat 1 tbsp of oil in the same wok. Stir-fry the shredded roast duck over low heat until fragrant. Add mashed plums, plum sauce and raw cane sugar. Bring to the boil. Put the prawns in and toss a few times. Add Chinese celery and toss again. Save on a serving plate. Garnish with sliced cucumber, coriander and red chillies.

Kitty's cooking tips

- Both fried prawns and roast duck are rich and flavourful. When put together in a dish, they tend to complement each other.
- You can use leftover roast duck for this dish. Don't be wasteful.

Assorted stir-fry with dried and fresh seafood

特色小炒王

準備時間

烹製時間

材料

魷魚鬚、鮮蝦仁各適量
蝦乾 40 克
豆豉鯪魚 1/2 條
銀魚仔 20 克
雞腿肉粒 40 克
鮮冬菇 2 朵
西芹 1 塊
紅辣椒 1 隻
脆腰果約 20 粒
四季豆 4 至 6 條
大豆芽 60 克
菜脯粒 1 湯匙
薑絲適量

調味料

XO 醬 1 茶匙
生抽 1/2 湯匙
糖 1/2 茶匙

準備

1. 魷魚鬚洗淨，切開；蝦仁開背、去腸，片開兩邊。加少許鹽和生粉拌勻。
2. 四季豆去頭尾，切段；鮮冬菇用濕布抹淨，切條。
3. 蝦乾浸軟；豆豉鯪魚瀝油，撕碎。
4. 西芹斜切片；紅辣椒去籽、切片；大豆芽切段。
5. 銀魚仔沖洗，吸乾水分，用中火油炸脆，瀝油。

做法

1. 用白鑊炒乾大豆芽，除去青味。
2. 燒熱油鑊，分別放入四季豆、冬菇、蝦乾、魷魚鬚和蝦仁走油。
3. 倒去炸油，留少許油在鑊底，燒熱油鑊，煎香雞肉，下豆豉鯪魚和薑絲一同炒透，再放入魷魚鬚、蝦仁、蝦乾、四季豆、西芹、紅辣椒、冬菇、大豆芽、菜脯粒，灒生抽，並加其他調味料一同炒勻，最後拌入脆腰果。
4. 上碟，放上炸銀魚仔，即可上桌。

三姐心得

■ 小炒王的用料很自由，喜歡食什麼就放什麼，例如傳統的小炒王有韮菜花，我改用四季豆、大豆芽和多種不同蔬菜，增加了甜味，配上鹹香的豆豉鯪魚和 XO 醬，味道取得平衡，豐富的材料令每一口都充滿不同的口感和食味，是一道下飯、配酒兩相宜的菜式。

Assorted stir-fry with dried and fresh seafood

- Preparation time: 20 minutes
- Cooking time: 10 minutes

Ingredients

squid tentacles
shelled shrimps
40 g large dried shrimps
1/2 canned dace with black beans
20 g dried anchovies
40 g boneless chicken thigh (diced)
2 fresh shiitake mushrooms
1 celery stalk
1 red chilli
20 cashew nuts
4 to 6 pods French beans
60 g soybean sprouts
1 tbsp diced salted radish
finely shredded ginger

Seasoning

1 tsp XO sauce
1/2 tbsp light soy sauce
1/2 tsp sugar

Preparations

1. Rinse the squid tentacles. Cut into pieces. Set aside. Cut along the back of the shelled shrimps. Devein and then cut into half along the back. Add a pinch of salt and caltrop starch. Mix well.
2. Snap off both ends of the French beans. Cut into segments and set aside. Wipe the shiitake mushrooms clean with a damp towel. Cut into strips.
3. Soak the dried shrimps in water till soft. Set aside. Drain the oil off the canned dace. Break it down into bits.
4. Slice the celery diagonally. De-seed and slice the red chilli. Cut soybean sprouts into short lengths.
5. Rinse the dried anchovies. Wipe dry. Deep-fry in oil over medium heat until crispy. Drain.

Method

1. Fry the soybean sprouts in a dry wok to remove the grass taste.
2. Heat oil in a wok. Blanch French beans, shiitake mushrooms, dried shrimps, squid tentacles and shelled shrimps in oil separately.
3. Drain most of the oil in the wok, saving a little. Then heat it up and fry the diced chicken until lightly browned. Add canned dace and shredded ginger. Toss well. Put in squid tentacles, shelled shrimps, dried shrimps, French beans, celery, red chilli, shiitake mushrooms, soybean sprouts and diced salted radish. Sprinkle with light soy sauce. Then add all remaining seasoning. Toss well. Stir in cashew nuts at last.
4. Save on a serving plate. Sprinkle with deep-fried anchovies on top. Serve.

Kitty's cooking tips

- You can use almost any ingredients for this recipe. Just put in whatever ingredients you like. Typically, this recipe calls for flowering chives. But this time, I use French beans, soybean sprouts and many other veggies to boost the sweetness, which balances with the savouriness of canned dace and XO sauce. The assortment fills each bite with different mouthfeels and flavours. It is truly a great dish to go with alcoholic drinks and steamed rice.

Braised fish fin with fresh and salted radish

鴛鴦蘿蔔燜魚天翅

準備時間 20 分鐘

烹製時間 20 分鐘

材料

龍躉魚天翅或頭腩 600 克
白蘿蔔 400 克
菜脯 150 克
枝竹 3 條
芹菜 2 條
大蒜 3 棵
葱段適量
指天椒 2 隻

調味料

腐乳 2 件
糖 1 湯匙
生抽 2 茶匙
鹽 1 茶匙
胡椒粉適量

準備

1. 魚天翅洗淨，吸乾水分，抹上少許鹽入味。
2. 白蘿蔔切滾刀段；菜脯縱切成長條；指天椒切段；芹菜拍扁，切斜段；大蒜切斜段。
3. 枝竹浸軟，切段。
4. 用白鑊慢火烘乾菜脯。

做法

1. 燒熱 3 湯匙油，下薑塊爆香，魚天翅拍上生粉，半煎炸至熟，取出，豎起放碟上。
2. 另起油鑊，下菜脯爆香，加入腐乳及白蘿蔔，加水 1 碗以中慢火燜 10分鐘，下枝竹、芹菜、大蒜、葱、指天椒、糖、鹽及生抽煮滾，再煮至收汁後，下胡椒粉拌勻。舀在魚天翅四周，即可上桌。

- 魚天翅是大型魚的背鰭。上碟時豎着擺放，象徵一帆風順，而且賣相十分有氣勢。

Braised fish fin with fresh and salted radish

■ Preparation time: 20 minutes
■ Cooking time: 20 minutes

Ingredients

600 g dorsal fin of giant grouper (or grouper head and belly)
400 g white radish
150 g salted radish
3 dried beancurd sticks
2 sprigs Chinese celery
3 leeks
spring onion sections
2 bird's eye chillies

Seasoning

2 cubes fermented beancurd
1 tbsp sugar
2 tsp light soy sauce
1 tsp salt
ground white pepper

Preparations

1. Rinse the fish. Wipe dry. Rub a pinch of salt evenly.
2. Cut the white radish into random wedges while rolling it on the chopping board. Set aside. Cut salted radish into long strips. Slice the bird's eye chillies. Crush the Chinese celery and slice diagonally. Slice the leeks diagonally.
3. Soak beancurd sticks in water till soft. Cut into short lengths.
4. Fry the salted radish in a dry wok over low heat until dry.

Method

1. Heat 3 tbsp of oil in a wok. Stir-fry ginger until fragrant. Coat the fish fin in caltrop starch lightly. Semi-deep fry until cooked through. Drain. Put the fish on a serving plate so that the fin stands up.
2. Heat another wok and add oil. Stir-fry salted radish until fragrant. Add fermented beancurd and white radish. Toss well. Add 1 bowl of water and bring to the boil. Turn to medium-low heat and simmer for 10 minutes. Add beancurd sticks, Chinese celery, leeks, spring onion, bird's eye chillies, sugar, salt and light soy sauce. Bring to the boil and cook until the sauce reduces. Sprinkle with ground white pepper and toss well. Pour this mixture into the serving plate around the fish. Serve.

Kitty's cooking tips

- Dorsal fin of giant grouper is believed to have lucky connotations of smooth sailing. The trick is to make the fin stands on the plate for a unique presentation with a strong presence.

Braised yellow croaker in lemon plum sauce

梅子
半煎煮黃花

準備時間
30
分鐘

烹製時間
20
分鐘

材料

大黃花魚 1 條
梅子 6 粒
香茅 1 支
蒜頭 2 至 3 瓣
薑 2 大片
檸檬 2 片
大蒜 3 棵
清水 1/2 碗

醃料

鹽 1 茶匙
胡椒粉適量
生粉 3 湯匙

調味料

魚露 1 茶匙
椰花糖或片糖碎 1 1/2 茶匙
檸檬汁 2 茶匙
糖 1 茶匙
薑汁 1 湯匙

準備

1. 黃花魚劏好洗淨，從魚肚片開，不要切斷魚脊，用 1 茶匙鹽及少許胡椒粉抹勻，醃半小時入味。
2. 薑連皮洗淨、拍裂；蒜頭去衣、切片；香茅用刀背拍裂，切段。
3. 梅子去核、壓碎；大蒜切斜段。

做法

1. 燒熱 4 湯匙油，黃花魚拍上生粉，下油鑊半煎炸至金黃色，盛起備用。
2. 另起油鑊，爆香薑片和香茅，下梅子、蒜片爆香，加入半碗清水煮滾，放入煎好的黃花魚，加入檸檬片和大蒜（喜歡吃辣的可加紅辣椒），下調味料同煮滾，轉慢火煮至收汁即可上碟。

- 半煎煮即先煎後煮，完成時不勾芡，連汁上碟。
- 椰花糖有椰子香味，增加菜式的味道層次。沒有的話，可用片糖代替。

Braised yellow croaker in lemon plum sauce

- Preparation time: 30 minutes
- Cooking time: 20 minutes

Ingredients

1 large yellow croaker
6 pickled plums in brine
1 stem lemongrass
2 to 3 cloves garlic
2 large slices ginger
2 slices lemon
3 leeks
1/2 bowl water

Marinade

1 tsp salt
ground white pepper
3 tbsp caltrop starch

Seasoning

1 tsp fish sauce
1 1/2 tsp coconut flower sugar (or crushed raw cane sugar)
2 tsp lemon juice
1 tsp sugar
1 tbsp ginger juice

Preparations

1. Dress the yellow croaker and rinse well. Cut it in half lengthwise along the belly without cutting all the way through to the dorsal side. Rub 1 tsp of salt and a pinch of ground white pepper all over the fish. Leave it for 30 minutes.
2. Rinse the ginger with skin on. Crush it with the flat side of a knife. Set aside. Peel and slice the garlic. Bruise the lemongrass with the back of a knife. Cut into short lengths.
3. Drain and de-seed the pickled plums. Mash with a fork. Set aside. Slice the leeks diagonally.

Method

1. Heat 4 tbsp of oil in a wok. Coat the fish lightly in caltrop starch. Fry in hot oil until golden on both sides. Set aside.
2. Heat another wok and add oil. Stir-fry ginger and lemongrass until fragrant. Add plums and garlic. Fry till fragrant. Add 1/2 bowl of water. Bring to the boil. Put in the fried fish. Add lemon and leeks (and optionally, red chilli). Add all seasoning and bring to the boil. Cook over low heat until the sauce reduces. Serve.

Kitty's cooking tips

- The fish is fried and then braised in the sauce, without adding a thickening glaze. Serve the fish with the thin sauce.
- Coconut flower sugar imparts a light nutty fragrant and gives the dish an extra dimension of flavour. If you can't find it, use raw cane sugar slab and crush it before using.

Braised fish head and swim bladders
with ginger and spring onion
薑葱焖
魚卜魚雲
準備時間
10
分鐘
烹製時間
20
分鐘

材料 

大魚頭 1 個
鯇魚卜 4 至 6 個
薑 100 克
葱 4 棵
紅葱頭 6 粒
陳皮絲少許

調味料 

生抽 2 湯匙
胡椒粉適量
蜆蚧醬 1 湯匙
紹酒 2 湯匙

準備 

1 魚頭請魚販代斬開兩邊，洗淨，瀝乾。

2 魚卜用刀切一個小口放氣，洗淨，瀝乾。

3 薑連皮拍碎；紅葱頭去衣，走油；葱切長段。

做法 

1 用一個大砂鍋加油燒熱，放入薑塊爆香；魚頭兩面都撲上生粉；魚雲向下，放入鍋中煎約 3 分鐘，至呈金黃色。

2 反轉魚頭，在鍋邊再加油，煎香另一面；並加入魚卜煎至金黃色，灒生抽。

3 放入紅葱頭，加約半碗熱水，煮滾後加蓋煮約 5 至 8 分鐘，待水分收乾，加葱段、蜆蚧醬及陳皮絲，再上蓋，約半分鐘後沿鍋蓋邊灒紹酒，陣陣酒香滲透鍋中即成，享用時灑上胡椒粉。

■ 自小就懂得「大魚頭，鯇魚尾」，只有大魚頭才有那兩塊白色半透明膠狀的魚雲，不要買錯其他的魚頭啊！

■ 魚卜必須戳穿，放出空氣才能煮熟。

Braised fish head and swim bladders with ginger and spring onion

- Preparation time: 10 minutes
- Cooking time: 20 minutes

Ingredients

1 head of bighead carp
4 to 6 swim bladders of grass carp
100 g ginger
4 sprigs spring onion
6 shallots
dried tangerine peel shreds

Seasoning

2 tbsp light soy sauce
ground white pepper
1 tbsp fermented clams
2 tbsp Shaoxing wine

Preparations

1. Ask the fishmonger to cut the fish head in half along the length. Rinse and drain well.
2. Rinse the swim bladders. Make a small cuts on each. Drain well.
3. Crush the ginger with skin on. Set aside. Peel the shallots. Deep fry for a while and set aside. Cut spring onion into long lengths.

Method

1. Heat oil in a large casserole pot. Put in the ginger and fry until fragrant. Coat both sides of the fish head in caltrop starch. Put it into the oil with the insides facing down. Fry for 3 minutes until golden.
2. Flip the fish head pieces. Add more oil along the rim of the casserole pot. Fry until golden. Add the swim bladders and fry till golden. Drizzle with light soy sauce.
3. Put in the shallots and add a half bowl of hot water. Bring to the boil. Cover the lid and cook for 5-8 minutes. When the water almost dry, put in the spring onions, fermented clams and dried tangerine peel shreds, cover the lid again for 30 seconds. Drizzle with Shaoxing wine along the rim of the lid and cook till you smell the winey fragrance. Sprinkle with pepper. Serve the whole pot.

Kitty's cooking tips

- I learned to use head of bighead carp, and tail of grass carp for cooking. Only bighead carp has the two globs of whitish translucent gelatinous structure in its head. Don't get other type of fish instead.
- You have to pierce the swim bladder and release the air before cooking it.

Braised grouper head and belly

脆燜斑頭腩

準備時間 20 分鐘

烹製時間 15 分鐘

材料

斑頭腩 1 斤
涼瓜 1/2 個
紅椒 1/2 隻，切絲
薑 3 片
蛋白 1 個
生粉適量

醃料

大地魚粉 1/2 茶匙
腐乳 2 件
薑汁 1/2 湯匙
生抽 1/2 湯匙
糖 1/2 茶匙
鹽 1/2 茶匙
胡椒粉少許
麻油少許

調味料

上湯 100 毫升
蠔油 1 湯匙
生粉水適量

準備

1. 斑頭腩斬碎，洗淨，瀝乾，用廚紙吸乾水分，下醃料（麻油除外）拌勻，最後下麻油和蛋白拌勻，醃片刻。
2. 涼瓜洗淨去瓤，斜切成塊。

做法

1. 斑頭腩撲上生粉，放入已燒至七分熱（約 200℃）的油中，大火炸至淺黃色，撈起；約 2 分鐘後回鑊再炸半分鐘，取出瀝油。
2. 另起油鑊爆香薑片，下涼瓜炒至變色，下蠔油炒勻，斑頭腩回鑊。
3. 上湯與生粉水混合，倒入鑊中煮滾，下紅椒絲，炒至斑頭腩及涼瓜掛芡，即可上碟。

■ 炸肉類和魚要用大火，才能鎖住水分；撈起後不要熄火，保持油溫，以便回鑊翻炸。翻炸後肉會夠熟，而且更香脆。

Braised grouper head and belly

- Preparation time: 20 minutes
- Cooking time: 15 minutes

Ingredients

600 g grouper head and belly
1/2 bitter melon
1/2 red chilli (shredded)
3 slices ginger
1 egg white
caltrop starch

Marinade

1/2 tsp ground dried plaice
2 cubes fermented beancurd
1/2 tbsp ginger juice
1/2 tbsp light soy sauce
1/2 tsp sugar
1/2 tsp salt
ground white pepper
sesame oil

Seasoning

100 ml stock
1 tbsp oyster sauce
caltrop starch slurry (1 part caltrop starch mixed with 1 part water)

Preparations

1. Chop the grouper head and belly into chunks. Rinse and drain. Wipe dry with paper towel. Add marinade (except sesame oil) toss well. Add egg white and sesame oil at last. Mix well and leave them briefly.
2. Rinse the bitter melon and remove the seeds. Cut into chunks diagonally.

Method

1. Coat the grouper pieces in caltrop starch. Heat enough oil for deep-frying up to 200°C. Deep-fry the grouper pieces until lightly browned. Remove from oil with a strainer ladle and leave them for about 2 minutes. Then put them back in the hot oil to fry for 30 seconds. Remove and drain excess oil.
2. In another work, heat some oil and stir-fry ginger until fragrant. Add bitter melon and stir-fry until it turns darker green. Add oyster sauce and toss well. Put the grouper pieces back in.
3. Mix the stock with the caltrop starch slurry. Pour into the wok and keep stirring until it boils. Add red chilli. Toss to coat the grouper and bitter melon evenly in the sauce. Serve.

Kitty's cooking tips

■ When you deep-fry meat or fish, make sure you do it over high heat to seal in the juices. When you remove the ingredients from the oil, do not turn off the heat. Let the oil heat up further before you put the ingredients back in. This two-step deep-frying method ensure the ingredients are cooked through while the crust will be crispy.

Fried crispy eel

吊燒脆鱔

準備時間 30分鐘

烹製時間 10分鐘

材料

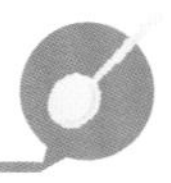

白鱔 1 段，約 500 克
雞蛋 1 個
麵粉 1 湯匙

醃料

鹽 1/2 茶匙
薑汁 2 茶匙
海鮮醬 1 茶匙
胡椒粉少許
生粉 1 湯匙

醬汁料

南乳 1 塊
糖 1 茶匙
熟油 1 茶匙

準備

白鱔請魚販劏好，去潺及起骨，洗淨，抹乾水分；用刀將魚肉㓤成鑽石紋，不要切斷魚皮（如㓤魷魚花），下醃料拌勻醃半小時。

做法

1. 將雞蛋掃勻白鱔上，再將整塊鱔蘸上麵粉。燒熱小半鑊油，用鐵勾吊起鱔塊，同時用勺不斷舀油淋在鱔上，一直淋至鱔熟透為止。如果覺得困難，可直接將鱔放入熱油中炸熟。
2. 炸好後用吸油紙吸去油分，切塊上碟；以南乳汁伴食。

- 鱔和其他魚不同，很耐火；而且必須熟透才好吃，未熟的鱔入口有辣味。
- 熟透的鱔會向外捲起，肉質變硬及有彈性。

Fried crispy eel

- Preparation time: 30 minutes
- Cooking time: 10 minutes

Ingredients

1 segment of white eel (about 500 g)
1 egg
1 tbsp plain flour

Marinade

1/2 tsp salt
2 tsp ginger juice
1 tsp Hoi Sin sauce
ground white pepper
1 tbsp caltrop starch

Dipping sauce

1 cube fermented tarocurd
1 tsp sugar
1 tsp cooked oil

Preparations

Ask the fishmonger to dress the eel, scrape off the slime and de-bone it for you. Rinse and wipe dry. On the meat side, make light crisscross incisions without cutting all the way through. Add marinade and mix well. Leave it for 30 minutes.

Method

1. Whisk the egg and brush it on the eel. Then coat the whole eel in plain flour. Heat less than half a wok of oil. Put the eel on a metal hook. Hold it by the hook over the wok of oil. Use a ladle to scoop hot oil and pour it over the eel on all sides until cooked through. If it's too difficult for you, put the eel into the hot oil and deep-fry until done.
2. Wipe off excess oil on the eel. Cut into pieces and save on a serving plate. Serve with the dipping sauce on the side.

Kitty's cooking tips

- As opposed to other fish, eel can stand prolonged cooking. In fact, it only tastes good if it's cooked long enough. Undercooked eel has a pungent taste which isn't too pleasing.
- When cooked through, the eel would curl outward while its flesh firms up and become springy.

Stir-fried eel in orange mango sauce

橙汁香芒炒鱔球

材料

白鱔 1 段，約 300 至 400 克
芒果 1 個
青、紅燈籠椒各 1/2 個
薑 2 片

醃料

鹽 1/2 茶匙
薑汁 2 茶匙
胡椒粉少許
生粉 1 湯匙

醬汁料

鮮橙汁 1 杯
糖適量
鹽少許
吉士粉水少許

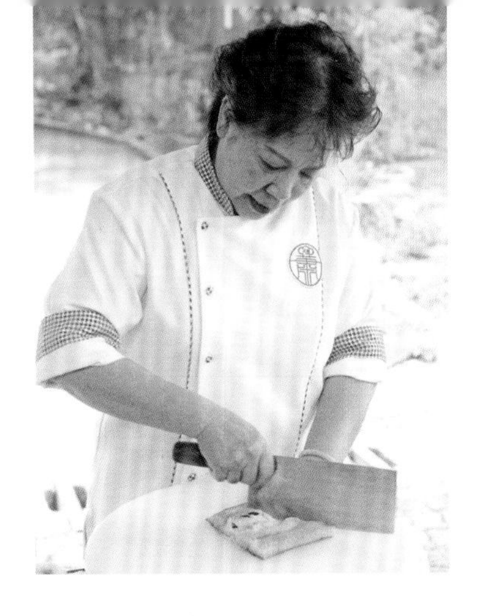

準備

1. 白鱔請魚販劏好、去潺及起骨，洗淨，抹乾水分；用刀將魚肉剞成鑽石紋，不要切斷魚皮（如切魷魚花），下醃料拌勻，醃半小時。
2. 芒果起肉、切塊；青、紅椒切塊。

做法

1. 燒熱油鑊，取出一勺熱油，將芒果浸熱。
2. 將薑放入油鑊中爆香，同時把鱔肉切角，下鑊炒至轉色和硬身，加入青、紅椒繼續炒。
3. 倒入橙汁，加鹽、糖調味，放入隔去油的芒果塊炒勻，下吉士粉水勾芡，即成。

三姐心得

- 橙和芒果的味道很夾，在橙汁裏加約 1/3 的芒果蓉可令果味更香濃。
- 用熱油浸芒果，熱力均勻地滲透芒果中，避免出現因炒的時間短而有外熱內冷的情況。

Stir-fried eel in orange mango sauce

- Preparation time: 30 minutes
- Cooking time: 8 minutes

Ingredients

1 segment of white eel (about 300 to 400 g)
1 mango
1/2 green bell pepper
1/2 red bell pepper
2 slices ginger

Marinade

1/2 tsp salt
2 tsp ginger juice
ground white pepper
1 tbsp caltrop starch

Sauce

1 cup freshly squeezed orange juice
sugar
salt
custard power slurry

Preparations

1. Ask the fishmonger to dress the eel, scrape off the slime and de-bone it for you. Rinse and wipe dry. On the meat side, make light crisscross incisions without cutting all the way through. Add marinade and mix well. Leave it for 30 minutes.
2. Skin and core the mango. Cut into pieces. Set aside. Cut bell peppers into chunks.

Method

1. Heat a wok and add 1 ladle of oil. Bring to the boil. Soak diced mango with 1 ladle of hot oil in a bowl. Drain well.
2. Stir-fry ginger in the wok until fragrant. Cut the eel into wedges. Put in the eel pieces and toss until they turn white and firm. Add bell peppers and keep stirring.
3. Add orange juice. Season with salt and sugar. Add diced mango and stir in the custard slurry. Cook till it thickens. Serve.

Kitty's cooking tips

- Orange and mango are the perfect match for each other. If you add 1/3 mango puree to the orange juice, the fruity flavour would be boosted further.
- I soak the diced mango in hot oil to make sure each dice is heated through properly. If you put in raw mango at last and toss it briefly, the core may still be cold.

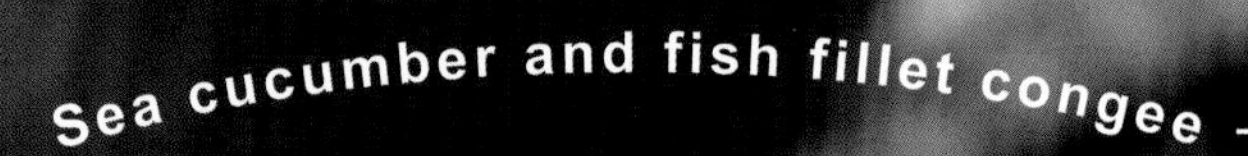

海參魚蓉粥

材料

魚 1 條，任何多肉的魚
有味粥底 1 湯碗
已浸發海參 1 條
嫩豆腐 1/2 件
新鮮粟米粒 1 湯匙
薑 2 片
芫荽碎少許

做法

1. 用砂鍋煮滾粥底，加入海參煮 5 分鐘，下粟米粒煮 2 分鐘，然後放入豆腐粒，煮滾後加入魚蓉拌勻，再煮 1 分鐘，盛入碗中供食。
2. 食時可撒上芫荽碎和胡椒粉。

準備

1. 魚劏好洗淨，吸乾水分，起出魚肉，抹上鹽和胡椒粉醃 1 小時；海參洗淨瀝乾，切粒。
2. 魚骨煎湯，隔渣，加米煲粥。
3. 起油鑊，爆香薑片，將魚煎至金黃色，取出，放涼，拆肉。
4. 豆腐切粒。

三姐心得

- 這個粥有魚肉的鮮味、海參和粟米粒的咬口，以及豆腐的幼滑，口感層次豐富，而且營養豐富，老幼咸宜。

Sea cucumber and fish fillet congee

- Preparation time: 30 minutes
- Cooking time: 10 minutes

Ingredients

1 fish (any fleshy fish)
1 large bowl seasoned congee base
1 sea cucumber (rehydrated)
1/2 cube soft tofu
1 tbsp fresh sweet corn kernels
2 slices ginger
finely chopped coriander

Preparations

1. Dress the fish and rinse well. Wipe dry. Skin the fish and debone. Rub salt and ground white pepper on the fish fillet. Leave it for 1 hour. Rinse the sea cucumber and drain well. Dice it.
2. Boil the fish bone in water. Strain the fish stock. Put rice into the fish stock and cook into congee until the rice grains are mushy.
3. Heat oil in a wok. Stir-fry ginger and fry the fish until golden. Set aside to let cool. Break into small bits and remove any bone remaining.
4. Dice the tofu.

Method

1. Boil the congee in casserole pot. Put in the sea cucumber and cook for 5 minutes. Add the sweet corn kernels and cook for 2 minutes. Put in the diced tofu and bring to the boil. Stir in the fish bits and mix well. Cook for 1 minute. Save into serving bowls and serve.
2. Sprinkle with coriander and ground white pepper before eating.

Kitty's cooking tips

- This congee has rich fish flavour. The sea cucumber and sweet corn give it a crunchy texture which sets against the velvety tofu nicely. Not only is the layering of mouthfeels amazing, it is also highly nutritious. It is a great staple for all ages.

Grouper fillet and salted fish fried rice in sweet corn sauce

粟米金銀魚燴飯

材料

石斑肉 150 克
鹹魚茸 40 克
紅葱頭 3 粒
雞蛋 2 個
有皮薑粒 2 湯匙
冷飯 1 大碗
罐頭粟米湯 200 克
葱粒 1 湯匙

醃料

糖 1/2 茶匙
鹽 1/2 茶匙
胡椒粉少許
蛋白 1/2 個

調味料

魚露 2 茶匙
鹽 1/4 茶匙
糖 1/2 湯匙
生粉水適量

準備

1. 石斑肉洗淨，用廚紙吸乾水分，切粒，拌入醃料醃 15 分鐘。
2. 紅葱頭去衣，切粒；薑連皮洗淨，切幼粒。

做法

1. 起油鑊，下薑粒和乾葱粒爆香，下鹹魚茸、雞蛋和飯，一同炒至飯粒會跳，下魚露及葱粒翻炒一下，上碟。
2. 另起油鑊，石斑肉撲上生粉，放入鑊中煎至金黃色，盛起；原鑊下粟米湯，放入鹽和糖調味，煮滾後用生粉水勾芡。
3. 斑塊放在炒飯上或沿碟邊圍成黃金圈，淋上粟米汁即可上桌。

- 薑爆香至中段時，下少量鹽令薑味更香。
- 炒飯前先洗濕雙手，抓鬆冷飯才下鑊，千萬不要用鑊鏟將飯戳碎。

Grouper fillet and salted fish fried rice in sweet corn sauce

- Preparation time: 20 minutes
- Cooking time: 15 minutes

Ingredients

150 g grouper fillet
40 g minced salted fish fillet
3 shallots
2 eggs
2 tbsp diced ginger (with skin on)
1 large bowl day-old rice
200 g canned cream style sweet corn
1 tbsp diced spring onion

Marinade

1/2 tsp sugar
1/2 tsp salt
ground white pepper
1/2 egg white

Seasoning

2 tsp fish sauce
1/4 tsp salt
1/2 tbsp sugar
caltrop starch slurry (1 part caltrop starch mixed with 1 part water)

Preparations

1. Rinse the grouper fillet. Wipe dry with paper towel. Dice it. Add marinade and mix well. Leave it for 15 minutes.
2. Peel the shallots and dice them. Rinse the ginger with skin on. Finely dice it.

Method

1. Heat a wok and add oil. Stir-fry diced ginger and shallot until fragrant. Add the minced salted fish, eggs and rice. Toss until the rice grains jump off the wok. Add fish sauce and spring onion. Toss a few times. Save on a serving plate.
2. Heat another wok and add oil. Coat the grouper fillet in caltrop starch. Fry in the oil until golden. Set aside. Pour in the cream style sweet corn. Add salt and sugar. Bring to the boil and stir in caltrop starch slurry. Cook till it thickens.
3. Arrange the grouper fillet over the fried rice at the centre or in a circle along the rim of the plate. Drizzle with the sweet corn sauce. Serve.

Kitty's cooking tips

- When you've fried the ginger half way through, sprinkle a pinch of salt to accentuate the aromas.
- For fried rice, wet your hands and squeeze the day-old rice in your hands to separate the grains before putting them in the wok. Do not break the chunks of rice with a spatula.

Steamed pork patty with diced pig ear

脆骨蒸肉餅

準備時間
5-60
分鐘

蒸製時間
35
分鐘

材料 

免治豬肉 220 克
滷豬耳 1/4 隻
馬蹄 3 至 4 顆
鹹蛋黃 1 個

調味料 

鹽 1/2 茶匙
魚露 1/2 茶匙
糖 1/2 茶匙
紹酒 1 茶匙
生粉 1 湯匙

準備 

1. 豬耳洗淨，用白滷水汁滷熟；或可買燒臘店的滷水豬耳。
2. 把豬耳切成綠豆般的小粒。
3. 馬蹄洗淨，去皮拍碎，粗剁成粒。

做法

1. 免治豬肉加入調味料，用手抓攪約 10 分鐘，再撻打至起膠。
2. 加入豬耳粒和馬蹄碎攪勻，然後搓圓成一個大肉球，放碟中，在頂部挖一個好像火山口的凹位，放上鹹蛋黃。
3. 放入蒸籠，以大火蒸 20 分鐘，即成有脆粒感的厚肉餅；取出，淋上少許生抽和麻油即可上桌。

三姐心得

■ 換一個食法：肉餅蒸好後，待稍涼，切成厚片，煎至兩面金黃上桌，味道更香濃，口感也更有層次。

Steamed pork patty with diced pig ear

- Preparation time: 5 to 60 minutes
- Cooking time: 35 minutes

Ingredients

220 g ground pork
1/4 marinated pig ear
3 to 4 water chestnuts
1 salted egg yolk

Seasoning

1/2 tsp salt
1/2 tsp fish sauce
1/2 tsp sugar
1 tsp Shaoxing wine
1 tbsp caltrop starch

Preparations

1. If you're using a fresh pig ear, rinse well and marinate it in a white spiced marinade until cooked through and flavourful. If you're using a store-bought marinated pig ear, skip this step.

2. Dice the pig ear finely (about the size of a mung bean).
3. Rinse the water chestnuts. Peel them and crush with the flat side of a knife. Then coarsely dice them.

Method

1. Add seasoning to the ground pork. Stir with your hand for about 10 minutes. Then lift the mixture off the bowl and slap it back in repeatedly until sticky.
2. Add diced pig ear and water chestnuts. Stir well. Then roll the mixture into a big ball. Put it into a steaming dish. Make an indentation at the centre and put in a salted egg yolk.
3. Transfer into a steamer and steam over high heat for 20 minutes. Remove from the steamer and drizzle with some light soy sauce and sesame oil. Serve.

Kitty's cooking tips

- For a variation, leave the pork patty to cool after steamed. Then slice thickly. Fry in a little oil until both sides golden and serve. That would boost the meaty flavour while giving an extra mouthfeel with the browned edges.

Coffee-scented pork ribs with peaches

蜜桃 咖啡肉排

準備時間 30 分鐘

烹製時間 15 分鐘

肉類

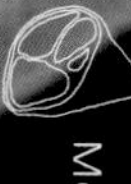

Meat

材料

腩排 300 克
罐頭蜜桃 4 件
生粉 150 克
雞蛋 1/2 個，拂勻

醃料

生抽 1/2 湯匙
鹽少許
生粉 1 茶匙
麻油少許
胡椒粉少許

咖啡汁

即溶咖啡粉 1 茶匙
淡奶 2 湯匙
生粉水 2 湯匙
蜜桃糖水 100 毫升
鹽 1/4 茶匙

準備

1. 腩排加醃料拌勻，醃 30 分鐘。
2. 蜜桃切成小塊。
3. 咖啡汁料調勻。

做法

1. 排骨下蛋汁拌勻，放入生粉用手輕抓，使所有排骨蘸滿生粉。逐少排骨放入熱油炸熟，撈起瀝油。不要關火，保持油溫；2 分鐘後把排骨回鑊用大火翻炸一下，即取出瀝油。
2. 倒去炸油，餘鑊底油，用中慢火下排骨及蜜桃炒熱，倒下咖啡汁不停翻炒，至汁全部裹住排骨，即可上碟。

- 為確保排骨熟透，炸時需將黏在一起的排骨用筷子分開。
- 留意炸第一次排骨時，應逐件放入鑊，避免食油濺起。

Coffee-scented pork ribs with peaches

- Preparation time: 30 minutes
- Cooking time: 15 minutes

Ingredients

300 g pork belly ribs
4 canned peach halves
150 g caltrop starch
1/2 egg (whisked)

Marinade

1/2 tbsp light soy sauce
salt
1 tsp caltrop starch
sesame oil
ground white pepper

Coffee sauce

1 tsp instant coffee granules
2 tbsp evaporated milk
2 tbsp caltrop starch slurry (1 part caltrop starch mixed with 1 part water)
100 ml syrup from canned peaches
1/4 tsp salt

Preparations

1. Add the marinade to the pork ribs. Mix well and leave them for 30 minutes.
2. Cut the peach halves into pieces.
3. Mix all coffee sauce ingredients together until well combined.

Method

1. Add the whisked egg to the pork ribs. Then put in caltrop starch and mix with your hands to coat them evenly. Put a few pieces of pork ribs at a time into the hot oil. Deep-fry until done. Remove from oil with a strainer ladle. Keep the heat on to maintain the temperature. Wait for 2 minutes. Put the pork ribs back into the hot oil all at once. Deep-fry over high heat briefly. Drain and set aside.
2. Drain off most oil in the wok, saving just a little. Heat the oil over medium-low heat. Put in the pork ribs and peach pieces. Add the coffee sauce and keep tossing until the sauce clings to the ribs evenly. Serve.

Kitty's cooking tips

- To make sure the pork ribs are cooked properly, separate the ribs that stick together with chopsticks when deep-frying them.
- When you fry the ribs the first time around, make sure you put in one by one to avoid the oil from splattering.

Pork ribs with pickled olive mustard greens and black bean sauce

欖菜豆豉骨

準備時間

烹製時間

材料

腩排 450 克
欖菜 4 湯匙
乾辣椒 15 克
蒜頭 2 瓣
豆豉 1 湯匙
炸蒜 1 湯匙
炸腰果及炸花生適量
生粉（炸排骨用）2 湯匙

醃料

生抽 1/2 湯匙
薑汁 1 湯匙
鹽 1/2 茶匙
糖 1/2 茶匙
紹酒少許

準備

1. 腩排洗淨、瀝乾，加醃料拌勻，醃 30 分鐘。
2. 蒜頭去衣，切成粗粒。

做法

1. 燒熱鑊加少許油，放入已隔去油分的欖菜，爆炒至香脆，倒在笊籬上，用鑊鏟將多餘的油壓走。
2. 鑊中加油 3 湯匙燒熱，排骨撲上生粉，逐少放入鑊中，半煎炸至乾身，取出，瀝油。
3. 倒去炸油，用鑊底餘油爆香蒜蓉和乾辣椒，將排骨回鑊，加欖菜和豆豉炒透，加入炸蒜、炸腰果及炸花生拌勻，即可上碟。

三姐心得

- 腩排要揀較薄身的，較嫩滑和易熟。
- 用熱油炸欖菜可逼出浸欖菜的油分，並令欖菜挺身，炒起來更香，口感也較好。

Pork ribs with pickled olive mustard greens and black bean sauce

- Preparation time: 30 minutes
- Cooking time: 15 minutes

Ingredients

450 g pork belly ribs
4 tbsp pickled olive mustard greens
15 g dried chilli
2 cloves garlic
1 tbsp fermented black beans
1 tbsp deep-fried garlic bits
deep-fried cashew nuts
deep-fried peanuts
2 tbsp caltrop starch (for coating the ribs)

Marinade

1/2 tbsp light soy sauce
1 tbsp ginger juice
1/2 tsp salt
1/2 tsp sugar
Shaoxing wine

Preparations

1. Rinse the pork ribs and drain well. Add marinade and mix well. Leave them for 30 minutes.
2. Peel the garlic. Coarsely dice it.

Method

1. Heat a wok and add oil. Drain excess oil off the pickled olive mustard greens. Fry them in a wok until crispy. Remove with a strainer ladle. Squeeze with a spatula to get rid of as much oil as you can.
2. In another wok, heat up 3 tbsp of oil. Coat the pork ribs in caltrop starch. Put the ribs in one by one and fry until dry on the surface. Remove from the oil and drain well.
3. Drain most oil from the wok, saving just a little. Stir-fry garlic and dried chilli until fragrant. Put the ribs back in and add pickled olive mustard greens and fermented black beans. Toss to heat through. Add deep-fried garlic bits, cashew nuts and peanuts. Toss well. Serve.

Kitty's cooking tips

- Pick pork belly ribs that are thinner in shape, so that they can be cooked more easily and stay juicy.
- Deep-frying the pickled olive mustard greens in oil helps drive the oil from inside the mustard greens. The mustard greens tend to be firmer in texture, with better mouthfeel and enhanced aromas.

準備時間
10
分鐘

烹調時間
2
小時

Beef shin cold appetizer with cucumber in Sichuan vinaigrette

酸汁凍牛腱伴青瓜

材料

牛腱 1 至 2 條，約 600 克
刺青瓜 2 條
芝麻少許

滷牛腱料

豉油雞汁 400 毫升
陳皮 2 塊
香葉 2 塊
沙薑 4 至 5 塊
桂皮 2 塊

汁材料

日本調味醋 3 湯匙
蒜蓉 1 湯匙
蜜糖 1/2 茶匙
指天椒 1 隻，切碎（隨意）
鹽 1/4 茶匙
花椒油 1 茶匙
麻油少許
香芹粒、芫荽碎各 1/2 湯匙

準備

1. 將豉油雞汁倒入大碗，加入香料蒸熱。
2. 牛腱解凍後洗淨，放入滾水煮 5 分鐘，去除雪氣及血水。
3. 上碟前 1 小時，將青瓜洗淨、切段，浸入加了少許白醋的冰水內。

做法

1. 將已氽水的牛腱放入上項滷水汁蒸 1.5 小時，以筷子能刺入為合；若未夠腍，再蒸 30 分鐘。撈出牛腱，涼後放入雪櫃至食用時取出。
2. 食用前，將青瓜段對切，再在皮上輕劃一兩刀，以乾淨毛巾蓋住（也可放在保鮮袋裏），用刀背拍裂，切成條，排放在碟上。
3. 牛腱切片放在青瓜上面。混合汁料，加入香芹粒和芫荽碎拌勻，淋於牛腱片上，再灑上芝麻即成。

三姐心得

- 因為青瓜是生吃的，浸時加些具殺菌功能的白醋更衛生。因已稀釋，青瓜不會帶酸味。
- 熟牛腱放入雪櫃後會變硬，所以必須蒸腍。蒸的時間視乎牛腱的大小，細的蒸 1 小時也可以。

Beef shin cold appetizer with cucumber in Sichuan vinaigrette

- Preparation time: 10 minutes
- Cooking time: 2 hours

Ingredients

1 to 2 pieces frozen beef shin (about 600 g)
2 prickly cucumbers
sesames

Spiced marinade

400 ml soy-based chicken marinade
2 pieces dried tangerine peel
2 bay leaves
4 to 5 slices sand ginger
2 pieces cassia bark

Sichuan vinaigrette

3 tbsp Japanese vinegar
1 tbsp grated garlic
1/2 tsp honey
1 bird's eye chilli (finely chopped, optional)
1/4 tsp salt
1 tsp Sichuan pepper oil
sesame oil
1/2 tbsp diced Chinese celery stems
1/2 tbsp finely chopped coriander

Preparations

1. Pour the soy-based chicken marinade into a big bowl. Add all spices and steam until hot.
2. Thaw the beef shin. Rinse well. Blanch in boiling water for 5 minutes to remove the blood and stale smell.
3. One hour before serving, rinse and cut cucumber into short lengths. Soak them in ice water with a dash of white vinegar added.

Method

1. Put the blanched beef shin into the chicken marinade. Steam for 1.5 hours until you can pierce it through with a chopstick. If it's not soft enough, steam for 30 more minutes. Remove the beef shin from the marinade and leave it to cool. Refrigerate until ready to serve.
2. Before serving, cut each short length of cucumber in half along the length. Then run a knife once or twice to scrape off part of the skin. Cover with a clean towel (or keep it in a zipper bag). Crush the cucumber with the back of a knife. Cut into strips. Arrange on a serving plate.
3. Slice the beef shin thinly and arrange over the cucumber. Mix the vinaigrette well. Add Chinese celery and coriander. Mix well and drizzle over the beef shin. Sprinkle with sesames on top. Serve

Kitty's cooking tips

- As the cucumber is served raw, I prefer dressing it with some white vinegar to kill the germs. But the white vinegar is diluted in water so that you won't taste it.
- After refrigerated, the beef shin will turn firmer. Therefore, it should be steamed till it feels soft enough when hot. Depending on its size, you should steam the beef shin for at least 1 hour.

Scrambled egg with pork balls, egg white omelette and spring onion

香蔥肉丸蛋中蛋

準備時間 5 分鐘

烹製時間 10 分鐘

材料 

雞蛋 5 個
蛋白 3 個
豬肉丸 4 粒
紅葱頭 3 粒
葱 3 條

準備 

1. 豬肉丸切片（每粒切成4至5片）。
2. 紅葱頭拍碎，粗剁幾下；葱切短段。
3. 蛋白拂勻；雞蛋打入大碗，不用攪散。

做法

1. 起油鑊，下蛋白煎至金黃色，盛起，切成塊件。
2. 再起油鑊，放下紅葱頭爆香；下肉丸片炒熱，然後加入蛋白塊。
3. 雞蛋加 1 茶匙油，用筷子輕輕拂勻，倒入鑊中，邊倒邊用筷子攪動（如黏鑊，沿鑊邊加少許油）。炒至雞蛋差不多凝固時，離火，撒上葱段，即可上碟。

三姐心得

■ 如果想更好味，可以將紅葱頭剁成蓉，用 1 湯匙油爆香 1 湯匙紅葱蓉，加入指天椒，離火。然後放 1 湯匙生抽、少許鮮醬油和糖在鑊內攪勻，做成香葱汁，淋上蛋面享用。

Scrambled egg with pork balls, egg white omelette and spring onion

- Preparation time: 5 minutes
- Cooking time: 10 minutes

Ingredients

5 eggs
3 egg whites
4 pork balls
3 shallots
3 sprigs spring onion

Preparations

1. Slice each pork balls into 4 to 5 slices.
2. Crush the shallots and coarsely chop a few time. Set aside. Cut spring onion into short lengths.
3. Whisk the egg whites. Set aside. Crack the eggs into a mixing bowl. Do not whisk them.

Method

1. Heat oil in a wok. Fry the egg white into an omelette until both sides golden. Set aside and cut into pieces.
2. Heat the wok again and add oil. Stir-fry the shallots until fragrant. Add sliced pork balls and toss until heated through. Put in the egg white omelette from step 1.
3. Add 1 tsp of oil to the eggs. Stir gently with chopsticks without mixing thoroughly. Pour into the wok. Stir with chopsticks. If it sticks to the wok, pour in a little oil along the rim of the wok. Stir until the egg is almost set. Turn off the heat and sprinkle with spring onion. Serve.

Kitty's cooking tips

- For a richer seasoning, finely chop the shallots. Stir-fry 1 tbsp of chopped shallot in 1 tbsp of oil until fragrant. Add chopped bird's eye chillies and turn off the heat. Add 1 tbsp of light soy sauce, a dash of Maggi's seasoning and a pinch of sugar. Stir well. This is the shallot sauce. Drizzle over the scrambled egg and serve.

準備時間 20 分鐘

烹製時間 1.5 小時

Beef brisket curry with sweet potato

番薯咖喱牛腩

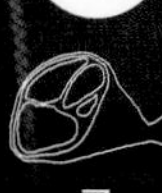

材料 

鮮牛腩 900 克
黃肉番薯 2 個
洋葱 1 個
番茄 1 個
香芹 1 棵
薑 2 塊

咖喱汁料

咖喱醬 2 至 3 湯匙（按個人口味）
花生醬 1 湯匙
淡奶 2 湯匙
椰漿 2 湯匙
鹽 1/2 茶匙
片糖碎 1 湯匙

蒸牛腩料 

香葉 3 至 4 片
桂皮 1 至 2 塊
沙薑 2 至 3 片
八角 3 至 4 粒
甘草 1 片
花椒 1 茶匙
冰糖 1 小粒
生薑 3 塊

準備

1. 牛腩原件洗淨，冷水下鍋，煮滾 3 至 5 分鐘，逌出血水；取出沖乾淨。
2. 番薯蒸熟，涼後去皮，切成大塊。
3. 洋葱、番茄、香芹切塊。

做法

1. 將已汆水的牛腩放蒸盆內，加蒸料及約 1 飯碗水，蒸 1 至 1.5 小時至腍（用筷子插入測試）。下鹽再蒸 10 分鐘，取出放涼後切件。蒸牛腩汁濾走香料及油，留用。
2. 起油鑊，慢火爆香薑，加入牛腩件爆至表面微黃，加洋葱和番茄略炒；加入咖喱醬和花生醬爆香，加水及蒸牛腩汁，下鹽和片糖調味，加蓋燜約 10 分鐘，加入番薯及淡奶再煮 1 至 2 分鐘，下香芹拌勻，若汁太稀可用生粉水勾芡。
3. 熄火，盛器內放入椰漿，倒入牛腩料，即可上桌。

三姐心得

- 牛腩以新鮮牛坑腩和挽手腩為上品。挽手腩位於爽腩及崩沙腩中間，相對其他部位有較多脂肪，特色是「一層脂肪一層肉」，每隻牛一般只有一斤半，是牛腩中的最上品。
- 煮咖喱時加入花生醬可提升味道層次。
- 待牛腩煮好後，加入椰漿不用煮，香味濃些。

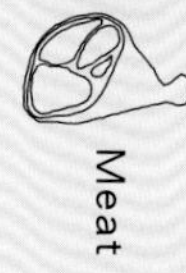

Beef brisket curry with sweet potato

- Preparation time: 20 minutes
- Cooking time: 1.5 hours

Ingredients

900 g freshly slaughtered beef brisket
2 yellow sweet potatoes
1 onion
1 tomato
1 sprig Chinese celery
2 slices ginger

Curry sauce

2 to 3 tbsp curry paste (or to taste)
1 tbsp peanut butter
2 tbsp evaporated milk
2 tbsp coconut milk
1/2 tsp salt
1 tbsp crushed raw cane sugar slab

Spices for steaming beef brisket

3 to 4 bay leaves
1 to 2 pieces cassia bark
2 to 3 slices sand ginger
3 to 4 pods star anise
1 slice liquorice
1 tsp Sichuan peppercorns
1 small piece rock sugar
3 slices fresh ginger

Preparations

1. Rinse the beef brisket. Put it in whole into a pot of cold water. Bring to the boil and cook for 3 to 5 minutes to remove the blood. Drain and rinse well.
2. Steam the sweet potatoes until tender. Let cool and peel them. Cut into chunks.
3. Cut onion, tomato and Chinese celery into pieces.

Method

1. Put the parboiled beef brisket into a steaming tray. Add spices and about 1 bowl of water. Steam for 60 to 90 minutes until you can poke it through with a chopstick easily. Season with salt and steam for 10 more minutes. Remove from steamer and let cool. Cut into chunks. Strain the juices and oil in the tray and set aside for later use.
2. Heat oil in a wok. Stir-fry ginger over low heat until fragrant. Put in the beef brisket and fry until lightly browned. Add onion and tomato. Toss briefly. Add curry paste and peanut butter. Stir until fragrant. Add water and the juices from steaming beef brisket. Season with salt and raw cane sugar slab. Cover the lid and cook for 10 minutes. Add sweet potatoes and evaporated milk. Cook for 1 to 2 minutes. Add Chinese celery and stir well. If the sauce is too thin, stir in caltrop starch slurry to thicken.
3. Turn off the heat. Pour coconut milk into a serving bowl. Pour in the beef brisket mixture. Serve.

Kitty's cooking tips

- For the best cuts of brisket, use fresh boneless short ribs or "handle" brisket as it is known in Hong Kong. "Handle" brisket is in fact the cut between the skirt and the hanger, with higher fat content than other cuts. It has alternate layers of fat and lean meat. Each cow only has about 900 g of "handle" brisket which is considered the most premium cut among all briskets.
- Adding peanut butter to curry gives it an extra dimension of flavours and depth.
- After beef brisket is cooked through, just pour it with the coconut milk. Coconut milk tastes richer uncooked.

Seared beef tenderloin cubes
with kohlrabi and glutinous rice cake
芥蘭頭
年糕牛柳粒
準備時間
20
分鐘
烹製時間
10
分鐘

材料

牛柳 300 克
芥蘭頭 1 個
上海年糕 150 克
青、紅椒各 1/2 個
洋葱 1/4 個
薑 2 塊
薑汁適量

醃料

油 1 湯匙
生粉 2 茶匙
糖 1 茶匙
生抽 1 茶匙

調味料

XO 醬 1 茶匙
蠔油 1 茶匙

準備

1. 牛柳洗淨，吸乾水分，切成骰子大的方粒。順序下油、生粉、糖和生抽（每下一種調味品拌勻後再下另一款），拌好，放置醃片刻。
2. 年糕切成約手指粗幼的長條，用熱水浸軟，瀝乾。
3. 芥蘭頭去皮，切片或切條；洋葱及青紅椒切塊。

做法

1. 燒熱油鑊，爆香薑塊，加入牛柳粒，大火炒至表面金黃，灒薑汁，離鑊。
2. 鑊內下少許油燒熱，加入芥蘭頭和年糕炒至變軟，加入 XO 醬及蠔油炒勻，將牛柳粒回鑊，翻炒一下即可上碟。

三姐心得

■ 可用西蘭花莖代替芥蘭頭。其實這兩種菜同屬十字科花植物，營養師認為兩種皆有很好的抗氧化功能，常吃有益。

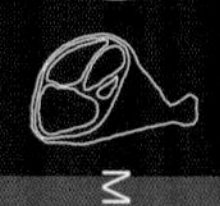

Seared beef tenderloin cubes with kohlrabi and glutinous rice cake

- Preparation time: 20 minutes
- Cooking time: 10 minutes

Ingredients

300 g beef tenderloin
1 kohlrabi
150 g Shanghainese glutinous rice cake
1/2 green bell pepper
1/2 red bell pepper
1/4 onion
2 slices ginger
ginger juice

Marinade

1 tbsp oil
2 tsp caltrop starch
1 tsp sugar
1 tsp light soy sauce

Seasoning

1 tsp XO sauce
1 tsp oyster sauce

Preparations

1. Rinse the beef and wipe dry. Cut into cubes about the size of a dice. Add oil, caltrop starch, sugar and light soy sauce in that particular order. Stir well after each addition. Set aside for a while.
2. Cut the glutinous rice cake into strips about the thickness of a finger. Soak it in hot water until soft. Drain.
3. Peel the kohlrabi. Slice or cut into strips. Set aside. Cut onion and bell peppers into chunks.

Method

1. Heat oil in a wok. Stir-fry ginger until fragrant. Put in the beef and fry over high heat until golden on all sides. Sprinkle with ginger juice. Set aside.
2. Heat a little oil in a wok. Put in the kohlrabi and glutinous rice cake. Toss until tender. Season with the seasoning. Put the beef back in. Toss a few times. Serve.

Kitty's cooking tips

- You can use broccoli stem instead of kohlrabi. Both of them belong to the brassicaceae family that nutritionists believe to have strong anti-oxidant effects. Frequent consumption may be beneficial to health.

準備時間
30
分鐘

烹製時間
10
分鐘

Pork spareribs in honey olive sauce

蜜汁欖角骨

材料

一字排 6 件
欖角 4 至 5 粒
蜜糖 2 湯匙
咕嚕汁 100 毫升
生粉（炸排骨用）2 湯匙

醃料

叉燒醬 1 茶匙
日本照燒汁 1 湯匙
磨豉醬 1 湯匙
薑汁 1 湯匙
紹酒少許

準備

1. 一字排洗淨、瀝乾，加醃料拌勻，醃 30 分鐘。
2. 欖角切碎（也可不切）。

做法

1. 鑊中加油 3 湯匙燒熱，排骨撲上生粉，放入鑊中半煎炸至熟，取出，瀝油。
2. 另起油鑊，放少量油，用慢火爆香欖角，加入蜜糖和咕嚕汁煮溶，放入排骨翻炒，至醬汁裹住排骨，即可上碟。

三姐心得

- 可以用牛仔骨代替一字排，做法一樣，只是醃牛仔骨時不要放鹽，而且煎至六成熟即可。
- 咕嚕汁材料及製法見第 60 頁。

Pork spareribs in honey olive sauce

- Preparation time: 30 minutes
- Cooking time: 10 minutes

Ingredients

6 pieces pork spareribs
4 to 5 Chinese preserved olives
2 tbsp honey
100 ml home-made sweet and sour sauce
2 tbsp caltrop starch (as deep-frying crust for spareribs)

Marinade

1 tsp Cha Siu sauce
1 tbsp teriyaki sauce
1 tbsp ground soybean paste
1 tbsp ginger juice
Shaoxing wine

Preparations

1. Rinse the spareribs and drain well. Add marinade and mix well. Leave them for 30 minutes.
2. Optionally, finely chop the preserved olives. (Or, you may leave them in whole.)

Method

1. Heat 3 tbsp of oil in a wok. Coat the spareribs in caltrop starch. Semi-deep fry in hot oil until cooked through. Drain and set aside.
2. Heat another wok again and add a little oil. Stir-fry preserved olives over low heat until fragrant. Add honey and sweet and sour sauce. Cook until honey dissolves. Put the spareribs back in and toss to coat evenly. Serve.

Kitty's cooking tips

- Instead of pork spareribs, you may use beef short ribs instead. The method is the same. Just make sure you don't add any salt to marinate the beef. Fry the short ribs in oil until medium in doneness. Do not overcook them.
- See p.61 for the recipe of the home-made sweet and sour sauce.

Stir-fried beef tenderloin with bitter melon and pickled ginger

涼瓜酸薑炒牛柳

準備時間 15 分鐘

烹製時間 8 分鐘

材料

牛柳邊 225 克
涼瓜 1/2 個，雷公鑿品種
酸薑 75 克
豆豉 1 湯匙
蒜蓉 1 湯匙
乾辣椒 20 克
薑 3 塊
薑汁適量
玫瑰露酒 1 湯匙

醃料

油 1 湯匙
生粉 2 茶匙
糖 1 茶匙
生抽 1 茶匙

準備

1. 牛柳邊洗淨，吸乾水分，逆紋切薄片。順序下油、生粉、糖和生抽（每下一種調味品拌勻後再下另一款），拌好，放置一旁醃片刻。
2. 涼瓜洗淨、抹乾，去瓤，縱切成粗絲。
3. 酸薑切塊。

做法

1. 燒熱油鑊，爆香薑塊，加入牛肉半煎 1 分鐘，反轉，濳薑汁，即取出。
2. 另起油鑊，下少許油燒熱，爆香蒜蓉和豆豉，加入乾辣椒和苦瓜翻炒，濳酒，下酸薑炒至苦瓜變軟，加入牛肉，略翻炒即可上碟。

三姐心得

- 這道菜的味道包含甜、酸、苦、辣，是三姐自創的獨特菜式。
- 牛柳邊是牛柳上半邊位，等於梅頭肉，脂肪多、肉味濃，有咬口又極鬆化，因為供應少，屬貴價部位。
- 炒牛肉必須將鑊預先燒到很熱。
- 自製酸薑做法：將 1.2 公升（兩瓶）米醋、450 克糖、1/3 茶匙味精及 1/2 茶匙鹽一同煮滾，放涼。子薑 1 斤（600 克）洗淨抹乾，刮皮，用粗鹽擦一下，待分泌水分，瀝乾。放入用滾水燙過晾乾的大玻璃瓶，注入糖醋，浸泡三數日可食用。

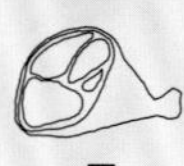

Stir-fried beef tenderloin with bitter melon and pickled ginger

- Preparation time: 15 minutes
- Cooking time: 8 minutes

Ingredients

225 g beef upper tenderloin
1/2 bitter melon (rotund-shaped)
75 g pickled ginger
1 tbsp fermented black beans
1 tbsp grated garlic
20 g dried chillies
3 slices ginger
ginger juice
1 tbsp Chinese rose wine

Marinade

1 tbsp oil
2 tsp caltrop starch
1 tsp sugar
1 tsp light soy sauce

Preparations

1. Rinse the beef. Wipe dry and slice thinly across the grain. Add oil, caltrop starch, sugar and light soy sauce in that particular order. Mix well after each addition. Set aside for a while.
2. Rinse the bitter melon. Wipe dry and de-seed it. Cut into thick strips along the length.
3. Cut the pickled ginger into pieces.

Method

1. Heat oil in a wok. Fry the ginger until fragrant. Put in the beef and fry for 1 minute. Flip and sprinkle with ginger juice. Set aside immediately.
2. Heat another wok and add some oil. Stir-fry garlic and fermented black beans until fragrant. Add dried chillies and bitter melon. Toss well. Sprinkle with Chinese rose wine. Add pickled ginger and keep tossing until the bitter melon tends soft. Put the beef back in. Toss a few times to mix well. Serve.

Kitty's cooking tips

- This recipe is the author's original creation, encompassing sweetness, sourness, bitterness and spiciness.
- Beef upper tenderloin is the beef equivalent of pork shoulder butt, with nice marbling of fat and robust meaty taste. It is soft, but not mushy. As it comes in strictly limited quantity, it is an expensive cut.
- Before you stir-fry beef in a wok, make sure you heat up the wok till very hot.
- You can also make your own pickled ginger from scratch. To make the pickling brine, put 1.2 litres of rice vinegar, 450 g white sugar, 1/3 tsp MSG and 1/2 tsp salt into a pot. Bring to the boil and leave it to cool. Wipe dry 600 g of young ginger. Scrape off the skin and rub coarse salt over it. Leave it briefly to draw the moisture out. Drain well. Put them into large glass jars that have been sterilized in boiling water and dried. Pour in the pickling brine to cover. Leave it for 3 days or so. Serve.

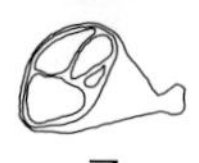

Chicken casserole with ginger, spring onion and fermented clams

薑葱蜆蚧雞煲

準備時間 10 分鐘

烹製時間 20 分鐘

材料

雞 1/2 隻
薑 3 至 4 塊
葱 4 棵，切長段
紅葱頭 6 粒
秋葵 4 條
紅辣椒 1 隻
陳皮絲少許

調味料

生抽 1 湯匙
蜆蚧醬 1 湯匙
紹酒 1 湯匙
糖 1/2 茶匙

準備

1. 雞洗淨，斬成小塊，加鹽 1/2 茶匙和生粉 1 茶匙拌勻入味。
2. 紅葱頭去衣，拍碎；秋葵洗淨，切蒂；紅辣椒洗淨，去籽、切片。

做法

1. 中火燒熱油鑊，下薑件爆香，放入雞件和紅葱頭爆至金黃色，有微焦香，加入秋葵和紅辣椒炒勻，灒生抽及下糖，加約 1/2 碗水，邊煮邊炒，把材料煮熟。
2. 待汁將收乾時，加入蜆蚧醬和陳皮絲炒勻煮滾，灒酒，轉入砂鍋，加葱段，即可上桌。

三姐心得

■ 除薑葱是必要材料外，可加入任何喜歡的蔬菜作為輔料。

Chicken casserole with ginger, spring onion and fermented clams

- Preparation time: 10 minutes
- Cooking time: 20 minutes

Ingredients

1/2 chicken
3 to 4 slices ginger
4 sprigs spring onion (cut into long segments)
6 shallots
4 okras
1 red chilli
shredded dried tangerine peel

Seasoning

1 tbsp light soy sauce
1 tbsp fermented clams
1 tbsp Shaoxing wine
1/2 tsp sugar

Preparations

1. Rinse the chicken and chop into pieces. Add 1/2 tsp of salt and 1 tsp of caltrop starch. Mix well.
2. Peel the shallots and crush them. Set aside. Rinse the okras and cut off the stems. Set aside. Rinse the red chilli. De-seed and slice it.

Method

1. Heat a wok over medium heat. Add oil. Stir-fry ginger until fragrant. Put in the chicken and shallots. Fry until lightly browned. Add okras and red chillies. Toss well. Sprinkle with light soy sauce and sugar. Add 1/2 bowl of water. Keep stirring while cooking until all ingredients are cooked through.
2. Cook till the sauce reduces. Add fermented clams and dried tangerine peel. Bring to the boil. Sprinkle with Shaoxing wine. Transfer into a casserole and put spring onion on top. Serve.

Kitty's cooking tips

- The ginger and spring onion are the irreplaceable key aromatics that give this dish its flavour profile. Other than that, you may add any veggies of your choice.

準備時間
20
分鐘
烹調時間
10
分鐘
Tom Yum lemon chicken wings
檸檬
冬蔭雞翼
家禽
Poultry

材料

雞中翼 6 隻
青檸檬 1 個
薑 4 片

醃料

鹽 1 茶匙
胡椒粉少許
生粉 2 茶匙

調味料

冬蔭功醬 1 湯匙
豉油雞汁 2 湯匙
玫瑰露酒 1/2 湯匙
蜜糖 2 湯匙

準備

1. 雞翼解凍洗淨，吸乾水分，用刀在雞翼背面兩條骨中間劃一刀，加鹽和胡椒粉拌勻醃半小時。
2. 檸檬榨汁，皮磨絲。

做法

1. 起油鑊，下薑爆香，下雞翼以中大火煎至金黃。
2. 放入冬蔭功醬炒一下，灒玫瑰露酒，下豉油雞汁和 2 湯匙檸檬汁，煮勻。最後收慢火，加入蜜糖翻炒，小心不要煮焦，至汁料全部掛上雞翼，即可上碟；上碟後撒檸檬皮絲裝飾。

■ 冬蔭功醬有香茅、南薑、辣椒、檸檬葉等香料，與檸檬很相配，輕微的辣味令雞翼的味道更豐富。

Tom Yum lemon chicken wings

- Preparation time: 20 minutes
- Cooking time: 10 minutes

Ingredients

6 mid-joint chicken wings
1 lime
4 slices ginger

Marinade

1 tsp salt
ground white pepper
2 tsp caltrop starch

Seasoning

1 tbsp Tom Yum paste
2 tbsp soy marinade for chicken
1/2 tbsp Chinese rose wine
2 tbsp honey

Preparations

1. Thaw the chicken wings and rinse well. Wipe dry. Make a cut on the back side of each wing between the two bones. Add salt and ground white pepper. Mix well and leave them for 30 minutes.
2. Squeeze the lime and grate the zest.

Method

1. Heat oil in a wok and stir-fry ginger till fragrant. Put in the chicken wings and fry over medium-high heat until golden.
2. Add Tom Yum paste and toss a few times. Sprinkle with Chinese rose wine. Add soy marinade for chicken and 2 tbsp of lime juice. Stir well. Turn to low heat and add honey. Toss until the sauce clings nicely to the wings. Make sure you don't burn the sauce. Transfer onto a serving plate. Sprinkle with grated lime zest at last. Serve.

Kitty's cooking tips

- Tom Yum paste is made of galangal, lemongrass, chillies and Kaffir lime leaves etc. It matches with citrus fruits such as lime perfectly. Its mildly spicy tastes add to the meaty flavours of the chicken wings.

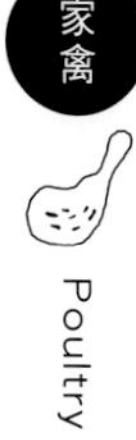

準備時間 20 分鐘

烹製時間 15-30 分鐘

Roast sesame squabs

芝麻乳鴿

材料

乳鴿 2 隻
滷水汁 250 毫升
清水 200 毫升
冰糖 1 小粒
蛋白 1 個
芝麻 1 湯匙

準備

1. 乳鴿清理內臟，斬去腳，洗淨瀝乾。
2. 滷水汁、水和冰糖煮滾，放入乳鴿煮 10 分鐘，期間反轉兩、三次，熄火，待在鍋 10 分鐘。
3. 取出乳鴿，吊起吹乾外皮。

做法

1. 在滷乳鴿皮掃上一層蛋白，黏上芝麻。
2. 焗爐預熱至中火（190℃ / 375 ℉），放入乳鴿烤焗至外皮金黃香脆。期間不時轉動，以免烤焦。

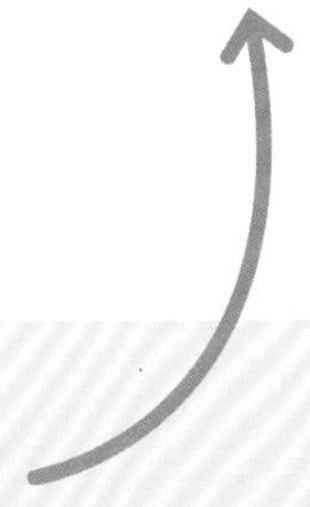

- 如沒有焗爐，可炸成紅燒乳鴿。先燒熱半鑊油，放入滷乳鴿炸 2 分鐘；用鋼勾針吊起乳鴿，淋油至皮脆且呈金黃色，瀝油，斬件上碟。
- 乳鴿肉細嫩，如煮過火，肉汁會流失，變得乾硬。
- 鴿皮吹至乾透，才能炸或烤得香脆。

Roast sesame squabs

- Preparation time: 20 minutes
- Cooking time: 15 to 30 minutes

Ingredients

2 squabs
250 ml spiced marinade
200 ml water
1 small cube rock sugar
1 egg white
1 tbsp sesames

Preparations

1. Dress the squabs and remove the innards. Cut off the feet. Rinse and drain well.
2. Put spiced marinade, water and rock sugar into a pot. Bring to the boil. Put in the squabs and cook for 10 minutes. Flip the squabs twice or three times throughout the cooking process. Turn off the heat and leave them in the marinade for 10 minutes.
3. Remove the squabs from the marinade. Hang in an airy spot and air-dry the skin.

Method

1. Brush a thin coat of whisked egg white over the squabs. Then coat them in sesames evenly.
2. Preheat an oven to 190°C (375°F). Bake the squabs until golden and crispy. Keep turning the squabs from time to time to avoid burning.

Kitty's cooking tips

- If you don't have an oven, you can deep-fry the squabs instead. Heat up half a wok of oil. Put in the squabs and deep-fry for 2 minutes. Then hang them on metal hooks. Hold them over the wok and use a ladle to pour hot oil over the squabs until golden. Drain and chop into pieces. Serve.
- Squabs have fine and delicate flesh. If overcooked, it turns dry and rubbery.
- You must air-dry the skin of the squabs well enough before frying or baking. Otherwise, the skin won't be crispy.

Minced dace patty with chives and lotus root

鯪魚肉韭菜煎藕餅

準備時間 20 分鐘

烹製時間 10 分鐘

材料

鯪魚肉 300 克
韭菜 100 克
蝦乾 40 克
馬蹄 4 顆
蓮藕 80 克
陳皮 1/2 塊

調味料

魚露 1/2 茶匙
鹽 1/4 茶匙
糖 1/4 茶匙
胡椒粉適量
生粉適量

蜆蚧蘸汁

蜆蚧醬 1 湯匙
蜆蚧汁 1 湯匙
玫瑰露酒 1/2 茶匙
陳皮絲少許
熱油 1/2 茶匙，後下

準備

1. 韭菜洗淨，切短段；馬蹄洗淨去皮，用刀拍碎；蓮藕切薄片，切粗粒。
2. 陳皮浸軟、去瓤，部分切幼絲，部分剁蓉。
3. 蝦乾浸軟，剁成蓉。

做法

1. 鯪魚肉、韭菜、馬蹄、蓮藕、蝦乾、陳皮蓉及調味料放入大湯碗，用手搓勻，順一方向不斷攪拌至起膠，再撻十數下。
2. 燒熱油鑊，將魚膠壓成厚圓餅狀，放入鑊以中慢火煎香並呈金黃色，反轉再煎，待至呈金黃色且魚肉熟透，切件上碟。
3. 品嚐時，蘸蜆蚧汁倍添美味。將陳皮絲和蜆蚧醬連汁放在小碟，加玫瑰露酒拌勻，灒 1/2 茶匙熱油，即成伴食蘸汁。

三姐心得

■ 這個魚餅放了蝦米和魚露，並加入大量脆口的蔬菜，吃起來鹹香惹味且較有嚼頭。

Minced dace patty with chives and lotus root

- Preparation time: 20 minutes
- Cooking time: 10 minutes

Ingredients

300 g minced dace
100 g Chinese chives
40 g dried shrimps
4 water chestnuts
80 g lotus root
1/2 piece dried tangerine peel

Seasoning

1/2 tsp fish sauce
1/4 tsp salt
1/4 tsp sugar
ground white pepper
caltrop starch

Fermented clam dipping sauce

1 tbsp fermented clams
1 tbsp juices from fermented clams
1/2 tsp Chinese rose wine
shredded dried tangerine peel
1/2 tsp sizzling hot oil (added last)

Preparations

1. Rinse the chives. Cut into short lengths. Rinse the peel the water chestnuts. Crush with the flat side of a knife. Set aside. Thinly slice the lotus root. Then coarsely chop it.
2. Soak dried tangerine peel in water till soft. Scrape off the white pith. Finely shred after of it. Then finely chop the other half.
3. Soak dried shrimps in water till soft. Finely chop them.

Method

1. Put all ingredients into a mixing bowl. Add seasoning. Stir with your hand in one direction until sticky. Lift the mixture off the bowl and slap it back in forcefully for over 10 times. That would make the dace mixture springier in texture.
2. Heat a wok and add oil. Press the dace mixture into a thick round patty. Fry it in the wok over medium-low heat until golden on one side. Flip it and fry till golden and cooked through. Slice and save on a serving plate.
3. To make the dipping sauce, put shredded dried tangerine peel, fermented clams and the juices into a small dish. Add Chinse rose wine and mix well. Pour in 1/2 tsp of sizzling hot oil. Stir and serve on the side.

Kitty's cooking tips

■ I put dried shrimps, fish sauce and lots of veggies into the minced dace, to give this dish a savoury richness and a crunchy mouthfeel.

Pan-fried stuffed okras

煎釀秋葵

準備時間 20 分鐘

烹製時間 10 分鐘

材料

秋葵 1 小包，約 10 條
墨魚膠 4 兩
綠色菜葉碎適量
紅辣椒粒少許，可選用
椒鹽少許

雞蛋漿

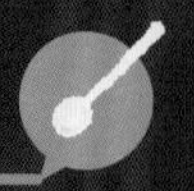

雞蛋 1 個
生粉 1 湯匙
油少許

準備

1. 秋葵洗淨，灼水 1 分鐘，抹乾水分。
2. 雞蛋漿材料拌勻，備用。

做法

1. 烹調前，將墨魚膠從雪櫃取出，混入菜葉碎和紅椒粒，用手攪勻再撻數下。墨魚膠包釀秋葵外，再蘸上雞蛋漿。
2. 起油鑊，油燒至六分熱（約 150℃），放入秋葵，半煎炸至金黃色，取出，撒下椒鹽，即可上碟。

三姐心得

■ 有些朋友怕秋葵滑潺潺的黏液而不吃。經過這樣處理，味道和質感完全改變，怕吃秋葵的人也會喜歡了；我的電視節目，有幾位工作人員就是實例。

Pan-fried stuffed okras

- Preparation time: 20 minutes
- Cooking time: 10 minutes

Ingredients

1 small pack okras (about 10)
150 g minced cuttlefish
finely chopped leafy greens (discard the stems)
diced red chilli (optional)
peppered salt

Deep-frying batter

1 egg
1 tbsp caltrop starch
oil

Preparations

1. Rinse the okras. Blanch in boiling water for 1 minute. Drain and wipe dry.
2. Mix the deep-frying batter ingredients together.

Method

1. Keep minced cuttlefish in the fridge before ready to use. Add leafy greens and diced red chilli to the minced cuttlefish. Stir with your hand. Lift it off the bowl and slap it back in forcefully for a few times. Wrap a thin layer of minced cuttlefish mixture on the okras. Dip them into the deep-frying batter.
2. Heat oil in a wok up to 150°C. Semi-deep fry the okras until golden. Drain. Sprinkle with peppered salt. Serve.

Kitty's cooking tips

- Some people may find okras repulsive because of the slime inside. In this recipe, the mouthfeel and taste of okras are completely altered. Even those averse to them will find this dish tasty. Some staffs of my TV show is a convert.

Stir-fried peanut sprouts with ground pork in XO sauce

XO 醬肉碎炒花生芽

準備時間 15 分鐘

烹製時間 10 分鐘

材料

花生芽 1 包
大豆芽 80 克
蝦米 1 湯匙
豬肉碎 100 克
青、紅辣椒各 1 隻
蒜蓉 1 湯匙

醃料

鹽 1/2 茶匙
生抽 1 茶匙
胡椒粉適量
麻油少許

調味料

XO 醬 1 湯匙
生抽 1 茶匙
薑汁 1 茶匙
鹽 1/2 茶匙
糖少許

準備

1. 豬肉加醃料拌勻。
2. 花生芽沖洗一下，瀝乾。
3. 大豆芽沖洗、瀝乾，切短段；辣椒洗淨，去籽、切絲。
4. 蝦米洗淨，浸軟。

做法

1. 燒熱白鑊，放入大豆芽炒香，取出。
2. 鑊內燒熱 2 湯匙油，放入花生芽走油約 3 分鐘，以去除草青味，撈出，瀝油。
3. 另起油鑊，豬肉碎炒熟，推至一邊。在鑊的另一邊放下蒜蓉爆香，與豬肉炒勻，加入 XO 醬及其他材料炒勻，灒生抽和薑汁，加鹽及糖調味，再炒 2 至 3 分鐘，至收汁即可上碟。

三姐心得

- 花生芽的白藜蘆醇含量比紅酒高得多，而且發芽還使花生中的蛋白質水解為氨基酸，有利於人體吸收，是近幾年新興的保健食材。生食熟食皆可。
- 走油令花生芽的花生粒炸至半熟，炒後才能熟透，花生不熟是不好吃的。

Stir-fried peanut sprouts with ground pork in XO sauce

- Preparation time: 15 minutes
- Cooking time: 10 minutes

Ingredients

1 pack peanut sprouts
80 g soybean sprouts
1 tbsp dried shrimps
100 g ground pork
1 green chilli
1 red chilli
1 tbsp grated garlic

Marinade

1/2 tsp salt
1 tsp light soy sauce
ground white pepper
sesame oil

Seasoning

1 tbsp XO sauce
1 tsp light soy sauce
1 tsp ginger juice
1/2 tsp salt
sugar

Preparations

1. Add marinade to the pork. Mix well.
2. Rinse the peanut sprouts. Drain.
3. Rinse the soybean sprouts. Drain and cut into short lengths. Set aside. Rinse the chillies. De-seed and finely shredded them.
4. Rinse the dried shrimps. Soak them in water till soft.

Method

1. Stir-fry the soybean sprouts without oil until fragrant. Dish up.
2. Heat a wok and add 2 tbsp of oil. Put in the peanut sprouts and gently move them around for 3 minutes. This helps remove the grassy taste. Drain and set aside.
3. Heat the wok again and add oil. Stir-fry ground pork till done. Push it to one side of the wok and put in the garlic. Stir-fry till fragrant. Toss the garlic and the pork together. Add XO sauce and the other ingredients. Toss well. Sprinkle with light soy sauce and ginger juice. Season with the salt and sugar. Toss for 2 or 3 more minutes until the sauce reduces. Serve.

Kitty's cooking tips

- Being a healthful ingredient in the past few years, peanut sprouts contain much more resveratrol than red wine. Germinating sprouts also hydrolyse the proteins in peanuts into amino acids, which are easier for human body to absorb. You can serve peanut sprouts raw or cooked.
- Blanching the peanut sprouts in oil helps deep-fry the peanut kernels on them till half-cooked. Then they can be fully cooked after stir-frying. Peanuts taste good only when fully cooked.

Steamed beancurd skin roll and angled luffa with ginger and garlic

薑蓉蒸勝瓜腐皮卷

材料 

勝瓜 1 至 2 條
鮮腐竹 1 包
紅棗 1 湯匙

調味料 

薑蓉 2 湯匙
鹽 1/2 茶匙
生抽少許

準備 

1. 勝瓜削去硬邊及部分皮，切成與腐竹相同約 5cm 段。
2. 紅棗洗淨、去核，切條。
3. 薑蓉加鹽拌勻，潛 1 茶匙熱油，成為油薑蓉。

做法 

1. 鮮腐竹分開，取半條捲起，用牙籤封口，與勝瓜段相間地排在蒸碟上。將油薑茸及紅棗鋪面。
2. 放上蒸籠，大火蒸約 5 分鐘，取出。淋少許生抽，即可供食。

三姐心得

- 這是一道簡單清爽而美觀的素菜，適合夏天食用。
- 勝瓜過生或過熟都不好吃，削皮多少也會影響口感。只要將棱邊刨平即可，不要刨掉太多皮。
- 可用杞子或金華火腿絲代替紅棗放面蒸。

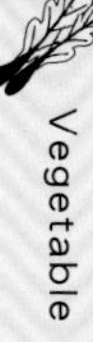

Steamed beancurd skin roll and angled luffa with ginger and garlic

- Preparation time: 10 minutes
- Cooking time: 8 minutes

Ingredients

1 to 2 angled luffa
1 pack fresh beancurd skin
1 tbsp red dates

Seasoning

2 tbsp grated ginger
1/2 tsp salt
light soy sauce

Preparations

1. Peel off the tough ridges on the angled luffa. Then peel off the skin on certain parts. Then cut into pieces about 5 cm long (similar to the width of the beancurd skin).
2. Rinse and de-seed red dates. Cut into strips.
3. Add salt to grated ginger. Mix well. Heat 1 tsp of oil until smoking. Drizzle on the ginger.

Method

1. Separate beancurd skin from each other. Take half of a sheet and roll it up. Secure the loose end with a toothpick. Arrange angled luffa and beancurd skin roll in alternate manner on a steaming dish. Evenly spread the grated ginger mixture and red dates all over.
2. Put the whole dish into a steamer. Steam over high heat for 5 minutes. Drizzle with some light soy sauce. Serve.

Kitty's cooking tips

- This is a simple, light and pretty dish for vegetarians. It is especially great for meals in summer.
- Angled Luffa doesn't taste good if it's undercooked or overcooked. Its mouthfeel depends on how much skin you peel off. Generally speaking, you just need to peel off the ridges. Don't peel off too much skin.
- You may finely shred Jinhua ham or dried goji berries instead of using red dates.

Assorted vegetables with ginger in sweet vinegar

尊貴素薑醋

準備時間 30 分鐘

烹製時間 2 小時

材料

老薑 600 克
白背木耳 20 克
番薯 300 克，品種不拘
黃耳 40 克
花生 75 克
黨參 1 大條
紅棗 12 顆
鹹蛋 3 個
雞蛋 3 個
甜醋 1.2 公升
黑米醋（酸醋）75 毫升

準備

1. 老薑洗淨，吹乾，刮去皮，在節上切開，用刀拍裂。番薯去皮，切件。
2. 白背木耳洗淨，用溫水浸軟，切塊及剪去硬蒂。
3. 黃耳洗淨，用溫水浸軟，汆水，放水喉下沖涼，切塊及剪去硬蒂。
4. 花生洗淨浸 4 小時，蒸腍。
5. 黨參、紅棗略沖洗；黨參切段；紅棗去核。
6. 雞蛋和鹹蛋烚熟，剝殼。

三姐心得

- 因為素薑醋沒有油，所以爆薑時要加點油。若煮豬腳薑醋，用白鑊爆薑即可。
- 第一次將兩塊黃耳加入醋內浸煮，黃耳會溶掉，令薑醋的膠質豐富。
- 不斷補充材料反覆煮，薑醋可保持達半年之久。

做法

1. 甜醋放入砂鍋煮滾，按個人口味分次下黑米醋調整味道。
2. 鑊內加約 1 湯匙油燒熱，下老薑爆炒至帶少許焦，連兩塊黃耳放入醋內，煮滾。
3. 加入花生、黨參、紅棗、木耳、鹹蛋和雞蛋，滾起後，轉慢火煲約 1 小時。
4. 放入番薯和其餘黃耳，煮滾後，轉慢火再煲 20 分鐘。熄火後不要開蓋，原煲放在陰涼地方，最少浸半天再煮滾，即可食用。

Assorted vegetables with ginger in sweet vinegar

- Preparation time: 30 minutes
- Cooking time: 2 hours

Ingredients

600 g old ginger
20 g wood ear fungus with pale underside
300 g sweet potatoes (of your choice)
40 g golden tremell (a.k.a. yellow fungus)
75 g peanuts
1 large piece Dang Shen
12 red dates
3 salted eggs
3 eggs
1.2 litres sweet vinegar
75 ml black rice vinegar

Preparations

1. Rinse the ginger and let it air-dry. Scrape off the skin. Cut along the nodes. Crack each segment with the flat side of a knife. Set aside. Peel the sweet potatoes and cut into chunks.
2. Rinse the wood ear fungus. Soak them in the warm water until soft. Cut into chunks. Trim off the hard stems.
3. Rinse golden tremell. Soak them in warm water until soft. Blanch in the boiling water. Rinse well under the tap water. Cut into chunks. Trim off the hard stems.
4. Rinse and soak peanuts in water for 4 hours. Steam them until tender.
5. Rinse Dang Shen and red dates. Cut Dang Shen into short lengths. De-seed the red dates.
6. Boil the eggs and salted eggs in water till cooked through. Shell them.

Method

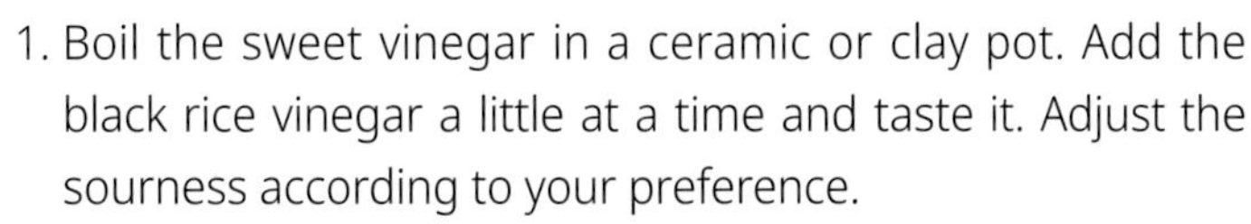

1. Boil the sweet vinegar in a ceramic or clay pot. Add the black rice vinegar a little at a time and taste it. Adjust the sourness according to your preference.
2. Heat 1 tbsp of oil in a wok. Put in the ginger and fry it till lightly browned. Transfer the ginger with 2 pieces of golden tremell into the vinegar mixture from step 1. Bring to the boil.
3. Add peanuts, Dang Shen, red dates, wood ear fungus, regular eggs and salted eggs. Bring to the boil. Turn to low heat and cook for 1 hour.
4. Put in sweet potatoes and the remaining golden tremell. Bring to the boil. Reduce to the low heat and cook for 20 more minutes. Cover the lid and turn off the heat. Leave the whole pot in a shaded cool place without opening the lid. Leave the whole pot for a half day. Reheat over a stove. Serve.

Kitty's cooking tips

- As opposed to the traditional pork trotter with ginger in sweet vinegar, this ovo-vegetarian recipe does not contain any meat. Therefore, the ginger should be fried in a little oil first. If you using pork trotter, fry the ginger in a dry wok without oil.
- At first, I put two pieces of golden tremell into the vinegar and cook them. The golden tremell will dissolve and make the vinegar more gelatinous.
- If you keep replenishing the ingredients and boil the mixture daily, the sweet vinegar can be kept as long as 6 months.

Deep-fried minced cuttlefish patty with Shanghainese Bok Choy

鍋貼小棠菜

準備時間
10
分鐘

烹製時間
10
分鐘

材料

小棠菜 2 至 3 棵
墨魚膠 300 克
海參粒 80 克
生粉適量

脆漿料

炸粉 4 至 5 湯匙
冷水 100 毫升
油 1 湯匙

蘸料

咕嚕汁適量，做法參考 p.60

準備

1. 小棠菜洗淨，抹乾水分，剁碎。
2. 炸粉用冷水調勻，至濃稠流質狀，靜置 20 分鐘，使用前加油 1 湯匙拌勻。

做法

1. 墨魚膠從雪櫃取出，混入海參粒和小棠菜碎，用手攪勻再撻至起膠。手沾油，將魚膠搓成圓球，壓扁，兩邊拍上生粉，裹上脆漿。
2. 燒熱 4 湯匙油，放入小棠菜墨魚餅，以中慢火半煎炸至兩面金黃香脆，取出，稍涼後切成方塊，上碟，伴咕嚕汁食用。

■ 這道菜有個故事：

二十多年前，三姐飯店主要經營火鍋，以每天新鮮手打丸子作賣點，曾有一位客人來要點獨特的菜式。當時店裏只有一些火鍋配料，大多數是手打丸。眼看手上有限的材料，我靈機一動，把墨魚肉加入剁碎的海參，加以調味，釀入小棠菜中，放入油鑊炸熟，配上糖醋汁及梅汁；那客人（實際是一個來挑機的廚師）吃了讚口不絕。自此，一道新菜式誕生了，彈牙加上香脆的口感，食客們都十分欣賞這道新菜式，曾創下日售六十碟的記錄。飯店因無法應付需求略為調整製作方法及形態，把小棠菜、墨魚膠及海參切碎並壓成餅狀，因此產生了「鍋貼小棠菜」。

Deep-fried minced cuttlefish patty with Shanghainese Bok Choy

- Preparation time: 10 minutes
- Cooking time: 10 minutes

Ingredients

2 to 3 sprigs Shanghainese Bok Choy
300 g minced cuttlefish
80 g diced sea cucumber
caltrop starch

Deep-frying batter

4 to 5 tbsp deep-frying batter mix
100 ml cold water
1 tbsp oil

Dipping sauce

home-made sweet and sour sauce (see method on p.61)

Preparations

1. Rinse the Shanghainese Bok Choy. Wipe dry. Finely chop it.
2. To make the deep-frying batter, mix cold water with deep-frying batter mix first until thickens. Leave it to stand for 20 minutes. Add 1 tbsp oil and stir well before cooking.

Method

1. In a mixing bowl, add diced sea cucumber and Shanghainese Bok Choy to the minced cuttlefish. Mix well. Then lift the mixture off the bowl and slap it back in forcefully until sticky. Grease your hands. Roll all cuttlefish mixture in your hands into a ball. Press gently into a patty. Spread with caltrop starch on both sides. Then dip it into the deep-frying batter.
2. Heat 4 tbsp of oil in a wok. Put in the cuttlefish patty. Semi-deep fry the patty over medium-low heat until both sides golden. Drain. Let cool briefly and cut into rectangular pieces. Serve with sweet and sour sauce.

Kitty's cooking tips

■ There's a story behind this dish. Over 20 years ago, my restaurant was selling hot pot dinner mainly with freshly hand-minced meatballs being the signature item. There was one customer who wanted to order something special, but hot pot ingredients such as meatballs were all I had in store. I took a good look at the ingredients and an idea flashed through my mind. I added diced sea cucumber into the minced cuttlefish, seasoned it and used it to stuff Shanghainese Bok Choy. I then deep-fried the stuffed greens and served it with sweet and sour sauce, and plum sauce. That customer (who was actually a chef coming to challenge me) was stunned and couldn't stop praising the dish. This has become another signature dish of my shop ever since – the springy cuttlefish works perfectly with the crispy crust. Diners couldn't get enough of this dish and it broke the record with 60 orders in one day. Later on, the kitchen really couldn't handle the complicated steps anymore, and had to fine-tune the method and the presentation. Therefore, instead of stuffing the Shanghainese Bok Choy, we finely chop it and mix it with the cuttlefish and sea cucumber. The mixture is shaped into a patty and fried.

Steamed eggplant with dried anchovies and chilli black bean sauce

銀魚乾
魚香蒸茄子

準備時間
5
分鐘

烹製時間
15
分鐘

材料

幼茄子 3 條
銀魚仔 40 克
老乾媽醬 2 茶匙
豆豉 1/2 湯匙

準備

1. 銀魚仔略為沖洗，瀝乾。
2. 茄子洗淨，縱切成四條，再切成幾段。

做法

1. 銀魚仔用熱油炸香，瀝油。
2. 將茄子排放在深碟內，排入炸銀魚仔，加入老乾媽醬和豆豉放面，以大火蒸 7 至 10 分鐘即可，吃時拌勻。

三姐心得

■ 茄子的品種很多，有長形的、蛋形的和圓形的，有深紫色、淺紫色、白色的，近年還有紫白間花皮的。蒸茄子宜選用幼長的品種，皮薄肉軟，味道較甜。

Steamed eggplant with dried anchovies and chilli black bean sauce

■ Preparation time: 5 minutes
■ Cooking time: 15 minutes

Ingredients

3 Chinese or Japanese eggplants
40 g dried anchovies
2 tsp Laoganma brand chilli sauce
1/2 tbsp fermented black beans

Preparations

1. Rinse the dried anchovies. Drain.
2. Rinse the eggplants. Cut along the length into quarters. Then cut into a few segments.

Method

1. Deep-fry the dried anchovies in hot oil until lightly browned. Drain.
2. Arrange the eggplants in a deep steaming dish. Put the deep-fried anchovies over the eggplants. Spread chilli sauce and fermented black beans all over. Steam over high heat for 7 to 10 minutes. Stir to mix well before serving.

Kitty's cooking tips

■ There are many varieties of eggplants – their shapes vary from slender to round and their colours vary from deep purple to light purple and white. There are even ones with mottled skin. For steaming, slender ones are preferred. Their skin is thin and their flesh is tender and sweet.

Assorted mushroom and rice cake omelette

雜菇蛋包年糕

菜蔬

Vegetable

準備時間 15 分鐘

烹製時間 10 分鐘

材料

上海年糕 150 克
雜菇 150 克
蝦仁 100 克
雞蛋 4 至 6 個
菜脯粒 1 湯匙
薑 2 片

調味料

鹽適量
胡椒粉適量
糖 1/2 茶匙
生粉水 1/2 湯匙

準備

1. 年糕切條，用熱水浸軟。
2. 雞蛋拂勻，下胡椒粉及鹽拌勻。
3. 雜菇依形狀切成幼絲或切粒。
4. 蝦仁洗淨，吸乾水分，切粗粒。

做法

1. 燒熱油鑊，爆香薑片，放入蝦仁炒香，下雜菇、年糕、菜脯粒同炒片刻，下糖調味，以生粉水勾芡，煮滾取出。
2. 鑊洗淨，加油燒熱，倒入蛋液及所有材料一齊煎 2 分鐘，翻至另一面煎片刻即成。

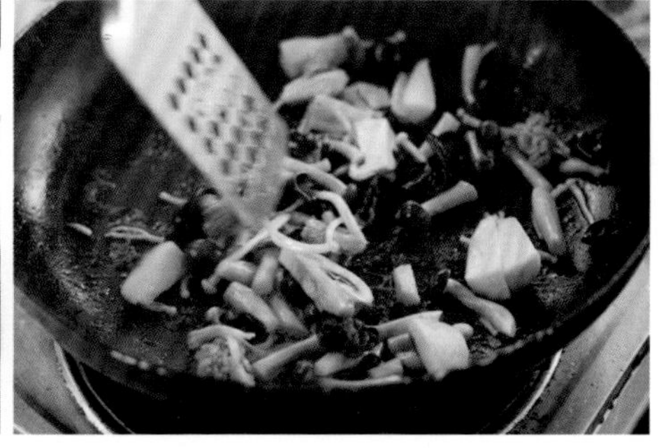

三姐心得

■ 無論哪種年糕，必須用滾水浸軟才煮，否則怎樣煮都不軟，影響口感。

Assorted mushroom and rice cake omelette

- Preparation time: 15 minutes
- Cooking time: 10 minutes

Ingredients

150 g Shanghainese glutinous rice cake
150 g assorted mushrooms
100 g shelled shrimps
4 to 6 eggs
1 tbsp diced dried radish
2 slices ginger

Seasoning

salt
ground white pepper
1/2 tsp sugar
1/2 tbsp caltrop starch slurry

Preparations

1. Cut glutinous rice cake into strips. Soak in hot water till soft.
2. Whisk the eggs. Season with salt and ground white pepper. Stir well.
3. Finely shred or dice the assorted mushrooms according to the shape.
4. Rinse the shrimps and wipe dry. Then cut them coarsely.

Method

1. Heat oil in a wok. Stir-fry ginger until fragrant. Stir-fry shelled shrimps until fragrant. Put in assorted mushrooms, glutinous rice cake and dried radish. Toss briefly. Season with the sugar. Stir in the caltrop starch slurry. Bring to the boil and set aside.
2. Wash the wok. Heat oil in a wok. Pour in the whisked eggs. Add all ingredients from step 1. Fry for 2 minutes. Flip and fry the other side until golden. Serve.

Kitty's cooking tips

- Whichever kind of rice cake you use, make sure you soak it in boiling water until soft before using. Otherwise, the rice cake won't be tender no matter how long you cook it.

Potato clam croquettes

蜆肉薯蓉餅

準備時間 20 分鐘

烹製時間 20 分鐘

材料

黃肉薯仔 2 個，約 300 克
蜆肉 100 克
洋葱 1/2 個
芹菜適量
生粉 2 茶匙
澄麵 1 湯匙
溶化牛油 1 湯匙

調味料

紅椒粉少許
糖 1/2 茶匙
鹽 1/2 茶匙

準備

1. 薯仔去皮、切片，蒸約 15 分鐘至軟，待稍涼，壓成薯蓉。
2. 洋葱去衣、切粒；芹菜切粒。
3. 蜆肉洗淨，吸乾水分。

做法

1. 將調味料及澄麵一起加入薯蓉攪勻，加入牛油搓至軟滑，然後下蜆肉、洋葱粒和芹菜粒搓勻。
2. 取小塊，搓圓，壓成厚餅，拍上生粉。
3. 起油鑊，放入薯餅煎至金黃，反轉，將各面及邊煎至金黃，即可上碟。

三姐心得

■ 黃肉薯仔質地粉甜，也較香，最適合做成薯蓉或薯餅。

Potato clam croquettes

- Preparation time: 20 minutes
- Cooking time: 20 minutes

Ingredients

2 yellow potatoes (about 300 g)
100 g shelled clams
1/2 onion
Chinese celery
2 tsp caltrop starch
1 tbsp potato starch
1 tbsp melted butter

Seasoning

paprika
1/2 tsp sugar
1/2 tsp salt

Preparations

1. Peel the potatoes and slice them. Steam for 15 minutes until tender. Leave them to cool briefly. Mash with a fork.
2. Peel and dice the onion. Set aside. Dice the Chinese celery.
3. Rinse the clams and wipe dry.

Method

1. Add seasoning and the potato starch into the mash potato. Mix well. Add butter and stir until smooth. Add clams, onion and Chinese celery. Knead to mix well.
2. Take some mash potato mixture in your hand. Roll into a ball. Then press into a thick patty. Coat it with caltrop starch. Repeat this step until all mash potato is used up.
3. Heat oil in a wok. Put in the croquettes and shallow-fry until golden. Flip them to fry the other side. Then fry the edges until golden. Serve.

Kitty's cooking tips

■ Yellow potato is starchy and flavourful. It is the best choice for making mash potato and croquettes.

Pan-fried tofu with Xue Cai
and enokitake mushrooms
雪菜金菇
煎豆腐
準備時間
15
分鐘
烹製時間
15
分鐘

材料 

布包豆腐 2 件
雪菜 100 克
金菇 1/2 包
毛豆 2 湯匙
薑 2 片
蒜蓉 1 茶匙
紅椒絲適量

調味料 

蠔油 1 湯匙
糖 1/2 茶匙
麻油少許
生粉水 1 茶匙

準備 

1. 毛豆加少許鹽煮熟或蒸熟。
2. 金菇切去根部，切成短段。
3. 雪菜洗淨，用清水浸淡，擠乾後切碎。
4. 豆腐用熱鹽水浸 15 分鐘，取出，一切開四，用毛巾吸乾水分，兩面抹上生粉。

做法 

1. 起油鑊，下油 2 湯匙燒熱，放入豆腐，以慢火煎 3 至 4 分鐘至表面金黃香脆，反轉再煎另一面。
2. 同時另起白鑊，炒乾雪菜，取出。另加 1 湯匙油燒熱，放入部分雪菜炸脆。
3. 取出雪菜，鑊底留少許油，下薑片及蒜蓉爆香，放入沒炸過的雪菜及金菇翻炒，加入蠔油、糖和麻油炒勻，加少量水煮片刻，下毛豆（愛吃辣的，可加入紅辣椒）炒勻，再煮片刻，下生粉水勾芡。
4. 將煎至皮脆肉嫩的豆腐排放上碟，淋上煮好的料頭，再放上已炸脆的雪菜和少許紅椒絲，即可上桌。

三姐心得

- 豆腐浸熱鹽水，會吸收鹹味和變得較結實，方便切開和煎脆。
- 這個食譜是一道素菜，喜歡的話，可加入剁碎的豬肉或雞肉同煮。

Pan-fried tofu with Xue Cai and enokitake mushrooms

- Preparation time: 15 minutes
- Cooking time: 15 minutes

Ingredients

2 cubes cloth-wrapped tofu
100 g Xue Cai (preserved potherb mustard)
1/2 pack enokitake mushrooms
2 tbsp young soybeans
2 slices ginger
1 tsp grated garlic
finely shredded red chilli

Seasoning

1 tbsp oyster sauce
1/2 tsp sugar
sesame oil
1 tsp caltrop starch slurry

Preparations

1. Sprinkle a pinch of salt on the young soybeans. Either blanch them in boiling water or steam them until cooked through.
2. Cut off the roots of the enokitake mushrooms. Cut into short lengths.
3. Rinse Xue Cai. Soak it in water to make it less salty. Squeeze dry and finely chop it.
4. Soak the tofu in hot salted water for 15 minutes. Cut into quarters. Wipe dry with towel. Coat both sides with caltrop starch.

Method

1. Heat a wok and add 2 tbsp of oil. Heat it up and fry the tofu over low heat for 3 to 4 minutes until crispy and golden. Flip it to fry the other side. Set aside.
2. Heat another dry wok. Stir-fry the Xue Cai until dry. Set aside. Add 1 tbsp of oil and heat it up. Put in half of the Xue Cai. Fry it until crispy. Set aside.
3. From the same wok, drain most of the oil, saving a little. Stir-fry ginger and garlic until fragrant. Put in the un-fried Xue Cai and enokitake mushrooms. Toss well. Add oyster sauce, sugar and sesame oil. Toss again. Add a little water and cook for a while. Put in the young soybeans. Optionally, put in red chilli at this point. Toss well and cook briefly. Stir in caltrop starch slurry.
4. Transfer the fried tofu on a serving plate. Pour the Xue Cai and enokitake mushroom glaze over. Top with deep-fried Xue Cai and shredded red chilli. Serve.

Kitty's cooking tips

- Soaking the tofu in hot salted water helps firm it up and pre-season it. You can slice it more easily and neatly. It also browns and turns crispy more easily.
- This is a vegetarian dish. For a meat variation, add ground pork or chicken to the glaze.

袋袋平安

Beancurd skin beggar's purse

準備時間 20分鐘

烹製時間 20分鐘

材料

日本豆卜 4 件或腐皮 1 張
韭菜 8 條
杞子適量
薑蓉 1 茶匙
蒜蓉 1 茶匙

餡料

馬蹄、芹菜、冬菇、雲耳、金菇、雪菜、紅蘿蔔各適量

調味料

麻油適量
糖 1 茶匙
紹酒 1 茶匙
蠔油 2 茶匙，分兩次使用
生粉水 2 湯匙，與半份蠔油拌勻

準備

1. 將所有餡料切幼粒。
2. 日本豆卜一剪為二，小心掀開。如用腐皮，剪成 8 份，修成圓形，用油炸脆，然後浸水，瀝乾備用。
3. 韭菜用熱水灼軟；杞子用熱水浸軟。

做法

1. 起油鑊，下蒜蓉爆香，盛起，放下所有餡料爆香，加入麻油、蠔油、糖，灒酒，盛起。
2. 將已爆香的餡料分成 8 份，用腐皮包起或放入豆卜袋中，紮成一小袋，用韭菜打結。
3. 放入碟上，蒸 3 分鐘。
4. 最後可放上杞子、薑蓉等增加風味，淋上蠔油芡即成。

三姐心得

■ 餡料要選用硬身的蔬菜，沙葛、馬蹄、毛豆、雲耳、芹菜等都適合；所有餡料宜切成幼粒。

Beancurd skin beggar's purse

- Preparation time: 20 minutes
- Cooking time: 20 minutes

Ingredients

4 Japanese tofu puffs (or 1 sheet beancurd skin)
8 sprigs Chinese chives
dried goji berries
1 tsp grated ginger
1 tsp grated garlic

Filling

water chestnuts
Chinese celery
dried shiitake mushrooms
cloud ear fungus
Enokitake mushrooms
Xue Cai (preserved potherb mustard)
carrot

Seasoning

sesame oil
1 tsp sugar
1 tsp Shaoxing wine
2 tsp oyster sauce (divided into 2 parts)
2 tbsp caltrop starch slurry (1 part caltrop starch mixed with 1 part water)

Preparations

1. Finely dice all filling ingredients.
2. Cut each tofu puff in half. Gently tear it apart to make a pouch. If you're using beancurd skin, cut it into 8 equal pieces. Trim into round discs. Deep-fry in oil until crispy. Then soak in water till soft. Drain well.
3. Blanch the Chinese chives in boiling water until soft. Drain. Soak goji berries in hot water until soft.

Method

1. Heat oil in a wok. Stir-fry diced garlic until fragrant. Set aside. Stir-fry all filling ingredients. Add sesame oil, 1 tsp of oyster sauce, sugar and Shaoxing wine. Toss and set aside.
2. Divide the filling into 8 equal parts. Fill each half of tofu puff with a part of filling, or wrap each part of filling in one piece of beancurd skin. Gather the edges upward into a packet. Tie the loose ends with a sprig of chives.
3. Arrange the packets neatly on a steaming plate. Steam for 3 minutes.
4. Arrange the goji berries and grated ginger over the beggar's purses. Heat 1 tsp of oyster sauce in a pan and stir in caltrop starch slurry. Pour the glaze over the beggar's purses. Serve.

 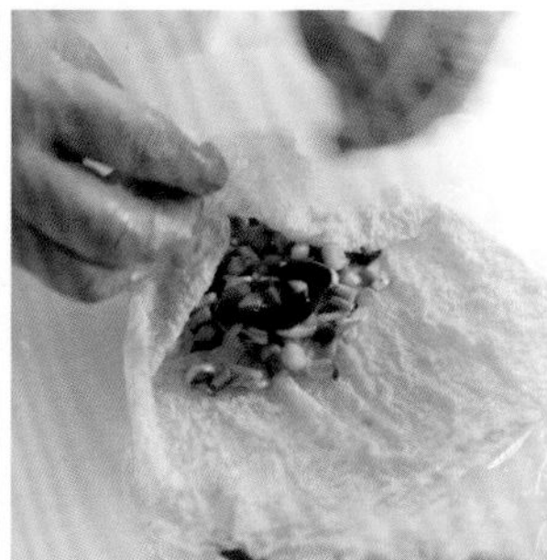 

Kitty's cooking tips

■ For the filling, feel free to use any veggie that is firm in texture after cooked, such as yam bean, water chestnuts, young soybeans, cloud ear fungus or Chinese celery. All filling ingredients should be finely diced.

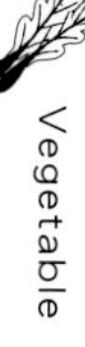

大廚小菜（增訂版）
三姐的50道創意家常菜

Master chef's home-style favourites (revised edition)
50 creative recipes by Chef Kitty Siu

作者
蕭秀香（三姐）

Author
Kitty Siu

策劃/編輯
簡詠怡

Project Editor
Karen Kan

攝影
梁細權

Photographer
Leung Sai Kuen

美術統籌及設計
鍾啟善

Design
Nora Chung

出版者
香港北角英皇道499號
北角工業大廈20樓
電話
傳真
電郵
網址

Publisher
Wan Li Book Company Limited
20/F., North Point Industrial Building,
499 King's Road, Hong Kong.
Tel: 2564 7511
Fax: 2565 5539
Email: info@wanlibk.com
Website: http://www.wanlibk.com
http://www.facebook.com/wanlibk

發行者
香港聯合書刊物流有限公司
香港荃灣德士古道220-248號
荃灣工業中心16樓
電話
傳真
電郵
網址

Distributor
SUP Publishing Logistics (HK) Ltd.
16/F., Tsuen Wan Industrial Centre, 220-248 Texaco Road,
Tsuen Wan, N.T., Hong Kong
Tel: 2150 2100
Fax: 2407 3062
Email: info@suplogistics.com.hk
Website: http://www.suplogistics.com.hk

承印者
中華商務彩色印刷有限公司
新界大埔汀麗路36號

Printer
C & C Offset Printing Co., Ltd.
36 Ting Lai Road, Tai Po, N.T., Hong Kong

出版日期
二零二二年五月第一次印刷
二零二三年六月第二次印刷

Publishing Date
First print in May 2022
Second print in June 2023

Published in Hong Kong, China.
Printed in China.
ISBN 978-962-14-7407-0